ASYMPTOTIC EXPANSIONS

A. ERDÉLYI
Professor of Mathematics
California Institute of Technology

DOVER PUBLICATIONS, INC., NEW YORK

This Dover edition, first published in 1956, is an unabridged and unaltered republication of Technical Report 3, prepared under contract Nonr-220 (11) for the Office of Naval Research Reference No. NR 043-121. It is published through special arrangement with the author.

Standard Book Number: 486-60318-0
Library of Congress Catalog Card Number: 56-14043

Manufactured in the United States of America
Dover Publications, Inc.
180 Varick Street
New York, N.Y. 10014

This booklet is based on a course of lectures given in the autumn of 1954 at the California Institute of Technology. The main purpose of the course was to introduce the students to various methods for the asymptotic evaluation of integrals containing a large parameter, and to the study of solutions of ordinary linear differential equations by means of asymptotic expansions. The choice of a comparatively wide range of topics resulted in a somewhat sketchy presentation; and the notes, having been prepared as the course progressed, lack finish. Nevertheless, the first printing was soon exhausted, and continued demand for copies seemed to warrant a second printing.

Chapter I contains a brief introduction to the general theory of asymptotic expansions. This subject is developed more or less to the extent to which it is needed to serve as a theoretical background for the main part of the course; and this modest chapter is by no means a substitute for the systematic and thorough study of the subject given in recent years by Professor van der Corput.

In Chapter II, the most important methods for the asymptotic expansion of functions defined by integrals are developed. This chapter owes much to the excellent pamphlet on this subject by Professor Copson. Lack of time precluded a very thorough discussion of any of the various methods (integration by parts, Laplace's method, method of steepest descents, method of stationary phase); and double and multiple integrals were left aside.

The remaining two chapters concern solutions of ordinary linear differential equations. In Chapter III, the "large" quantity is the independent variable in the differential equation. The discussion is restricted to a differential equation of the second order for which infinity is an irregular singular point of rank one with a characteristic equation which has two distinct roots. In Chapter IV, the "large" quantity is a parameter in the differential equation. The discussion is restricted to a differential equation of the second order with a real independent variable ranging over a bounded closed interval. Both Liouville's approximation and its generalization appropriate for an interval containing a transition point are given.

The preparation of these notes was supported by the Office of Naval Research. The author wishes to acknowledge his thanks to Mr. C.A. Swanson for very able and valuable assistance rendered in the course of preparation of the notes.

December 1955. A. ERDÉLYI

CONTENTS

CHAPTER III

SINGULARITIES OF DIFFERENTIAL EQUATIONS

CHAPTER IV

DIFFERENTIAL EQUATIONS WITH A LARGE PARAMETER

INTRODUCTION

It happens frequently that a divergent infinite series may be used for the numerical computation of a quantity which in some sense can be regarded as the "sum" of the series. The typical situation is that of a series of variable terms whose "sum" is a function, and the approximation afforded by the first few terms of the series is the better the closer the independent variable approaches a limiting value (often ∞). In most cases the terms of the series at first decrease rapidly (the more rapidly the closer the independent variable approaches its limiting value) but later the terms start increasing again. Such series used to be called *semi-convergent* (Stieltjes), and numerical computers often talk of *convergently beginning* series (Emde); but in the mathematical literature the term *asymptotic series* (Poincaré) is now generally used. We shall see later that asymptotic series may be convergent or divergent.

Let us consider an example first discussed by Euler (1754). The series

$$(1) \quad S(x) = 1 - 1!\,x + 2!\,x^2 - 3!\,x^3 + \cdots = \sum_0^\infty (-1)^n n!\,x^n$$

is certainly divergent for all $x \neq 0$, yet for small x (say 10^{-2}) the terms of the series at first decrease quite rapidly, and an approximate numerical value of $S(x)$ may be computed. What function of x does this numerical value represent approximately?

Euler considers $\phi(x) = x\,S(x)$. Then

$$\phi'(x) = 1! - 2!\,x + 3!\,x^2 - \cdots = \frac{x - \phi(x)}{x^2},$$

or

$$x^2\,\phi'(x) + \phi(x) = x,$$

and $\phi(x)$ may be obtained as that solution of this differential equation which vanishes as $x = 0$. Alternatively, we use Euler's integral of the second kind,

$$n\,! = \int_0^\infty e^{-t} t^n \, dt$$

and obtain

$$S(x) = \int_0^\infty e^{-t} dt - x \int_0^\infty e^{-t} t \, dt + x^2 \int_0^\infty e^{-t} t^2 \, dt - \cdots$$

$$= \sum_0^\infty (-1)^n \int_0^\infty e^{-t} (xt)^n \, dt.$$

If we formally sum under the integral sign, $S(x)$ becomes

$$(2) \quad \int_0^\infty \frac{e^{-t}}{1+xt} \, dt.$$

Now,

$$(3) \quad f(x) = \int_0^\infty \frac{e^{-t}}{1+xt} \, dt$$

is a well-defined function of x, as a matter of fact an analytic function of x in the complex x-plane cut along the negative real axis, and it is closely related to the so-called exponential integral. The question then arises: in what sense does the divergent series (1) represent the function (3)? To answer this question, we note that for $m = 0, 1, 2, \ldots$

$$\frac{1}{1+xt} = \sum_{n=0}^{m} (-xt)^n + \frac{(-xt)^{m+1}}{1+xt}$$

and hence

$$(4) \quad f(x) = S_m(x) + R_m(x)$$

where

$$(5) \quad S_m(x) = \sum_{n=0}^{m} (-1)^n n!\, x^n$$

is a partial sum of (1), and

$$(6)\quad R_m(x) = (-x)^{m+1} \int_0^\infty \frac{e^{-t}\, t^{m+1}}{1 + xt}\, dt$$

is the remainder.

If Re $x \geq 0$, we have $|1 + xt|^{-1} \leq 1$ and

$$(7)\quad |R_m(x)| \leq (m+1)!\, |x|^{m+1} \qquad \text{Re } x \geq 0.$$

On the other hand, if Re $x < 0$, $\phi = \arg x$, and $\pi/2 < \pm\phi < \pi$, then

$$|1 + xt|^{-1} \leq |\operatorname{cosec} \phi|,$$

and

$$(8)\quad |R_m(x)| \leq (m+1)!\, |x|^{m+1}\, |\operatorname{cosec} \phi|, \qquad \text{Re } x < 0.$$

In either case, the remainder is of the order of the first "neglected" term of $S(x)$, and approaches 0 rapidly as $x \to 0$. The limit is uniform in any sector $|\arg x| \leq \pi - \epsilon$, $\epsilon > 0$. If Re $x \geq 0$, the remainder is numerically less than the first neglected term, and if $x > 0$, the remainder has also the sign of the first neglected term. Thus, for $x > 0$, the series (1) behaves very much like a convergent alternating series, except that the smallest term of (1), which occurs when m is approximately equal to x, determines a limit to the accuracy beyond which it is impossible to penetrate.

The theory of asymptotic series was initiated by Stieltjes (1886) and Poincaré (1886). We may distinguish two parts of the theory. One part, which we may call the theory of asymptotic *series*, treats topics such as "sums" of asymptotic series ("asymptotic limits", "asymptotic convergence"), and operations with asymptotic series (algebraic operations, differentiation, integration, substitution of asymptotic expansions of a variable in convergent or asymptotic series involving this variable, and the like). The most comprehensive presentation of this part of the theory is to be found in van der Corput's Lectures (1951, 1952) and current publications by the same author. In these pages we shall restrict ourselves to a brief introduction to the theory of asymptotic series, and shall devote most of our attention to the other part of our subject, to the theory of asymptotic *expansions*. Here the central theme is the construction and investigation of series which represent given functions asymptotically. The functions are often given by integral representations, or by power series, or else appear as solutions of differential equations; and in the latter case the "variable" of the asymptotic expansions may occur either as the independent variable, or else as a parameter, in the differential equation.

REFERENCES

van der Corput, J.G., 1951: *Asymptotic expansions*, Parts I and II. National Bureau of Standards (Working Paper).

van der Corput, J.G., 1952: *Asymptotic expansions*, Part III. National Bureau of Standards (Working Paper).

van der Corput, J.G., 1954a: *Nederl. Akad. Wetensch., Amsterdam, Proc.* 57, 206-217.

van der Corput, J.G., 1954b: *Asymptotic Expansions* I. *Fundamental theorems of Asymptotics*. Department of Mathematics, University of California, Berkeley.

Euler, Leonhard, 1754: *Novi commentarii ac. sci. Petropolitanae* 5, 205-237. *Opera omnia*, ser. I, 14, 585-617, in particular, 601ff.

Poincaré, H., 1886: *Acta Math.* 8, 295-344.

Stieltjes, Th., 1886: *Ann. de l'Éc. Norm. Sup.* (3) 3, 201-258.

CHAPTER I

ASYMPTOTIC SERIES

1.1. O-symbols

In general, the "independent variable" will be a real or complex variable, but in this chapter x stands for a variable element of a topological T_2-space (Hausdorff space) except when stated otherwise. The variable x ranges over a set R, and x_0 is a limit point of R (which may or may not belong to R). $\phi(x)$, $\psi(x)$, and similar symbols denote real- or complex-valued numerical functions of x defined when x is in R.

The following *order relations* involving the *order symbols* O, o will be used. We write $\phi = O(\psi)$ in R if there exists a constant (i.e., number independent of x) A so that $|\phi| \leq A|\psi|$ for all x in R; $\phi = O(\psi)$ as $x \to x_0$ if there exists a constant A and a neighborhood U of x_0 so that $|\phi| \leq A|\psi|$ for all x common to U and R; and we write $\phi = o(\psi)$ as $x \to x_0$ if for any given $\epsilon > 0$ there exists a neighborhood U_ϵ of x_0 so that $|\phi| \leq \epsilon|\psi|$ for all x common to U_ϵ and R. If $\psi \neq 0$ in R then the three conditions may be formulated more simply: $\phi = O(\psi)$ in R [as $x \to x_0$ in R] if ϕ/ψ is bounded in R [as $x \to x_0$ in R], and $\phi = o(\psi)$ as $x \to x_0$ if $\phi/\psi \to 0$ as $x \to x_0$.

In the following examples x is a complex variable, and S_Δ is the sector $0 < |x| < \infty$, $|\arg x| < \pi/2 - \Delta$. The reader should verify each assertion. (i) $e^{-x} = O(x^\alpha)$, $e^{-x} = o(x^\alpha)$ as $x \to \infty$ in S_Δ, $\Delta > 0$, α arbitrary; and neither of these order relations holds (for arbitrary α) when $\Delta \leq 0$. (ii) $e^{-x} = O(x^\alpha)$ as $x \to \infty$ in S_0 provided that Re $\alpha \geq 0$; this order relation fails to hold when Re $\alpha < 0$. (iii) $e^{-x} = O(x^\alpha)$ in S_Δ provided that either $\Delta > 0$ and Re $\alpha \leq 0$ or $\Delta = 0$ and Re $\alpha = 0$.

If the functions involved in an order relation depend on parameters, in general also the constant A, and the neighborhoods U, U_ϵ involved in the definitions will depend on the parameters. If A, U, U_ϵ may be chosen to be independent of the parameters, the order relation is said to hold *uniformly* in the parameters.

Operations with order relations are governed by a number of simple rules. We shall set out the more frequently used rules for the O-symbol: the corresponding rules hold for the o-symbol. In the following rules R and x_0 are fixed, and the qualifying phrase "as $x \to x_0$" is omitted throughout.

If $\phi = O(\psi)$ and $a > 0$ then

(1) $|\phi|^a = O(|\psi|^a)$.

If $\phi_i = O(\psi_i)$, $i = 1, \ldots, k$ and the a_i are constants, then

(2) $\sum_i a_i \phi_i = O(\Sigma |a_i|\, |\psi_i|)$.

This relation holds also for infinite series provided that $\phi_i = O(\psi_i)$ uniformly in i. In the case of infinite series, equation (2) and similar statements will be interpreted in the following manner. If $\Sigma |a_i \psi_i|$ converges then so does $\Sigma\, a_i \phi_i$ and (2) is true, and if $\Sigma\, |a_i \psi_i|$ diverges then there is nothing to state.

If $\phi_i = O(\psi_i)$, $i = 1, \ldots, k$, the a_i are constants, and $|\psi_i| \leq \psi$ for $i = 1, \ldots, k$ and for all x common to R and to some neighborhood U_0 of x_0, then

(3) $\sum_i a_i \phi_i = O(\psi)$.

This relation holds for infinite series provided that $\phi_i = O(\psi_i)$ uniformly in i, and $\Sigma\, |a_i| < \infty$.

If $\phi_i = O(\psi_i)$, $i = 1, \ldots, k$ then

(4) $\prod_i \phi_i = O(\prod_i \psi_i)$.

The proof of (1) is immediate. To prove (2), we remark that by assumption there are numbers A_i and neighborhoods U_i of x_0 associated with the ϕ_i. If the number of the ϕ_i is finite, there is an A larger than all the A_i, and a neighborhood U contained in all the U_i, and

$$|\Sigma\, a_i \phi_i| \leq \Sigma\, |a_i|\, A_i |\psi_i| \leq A \Sigma\, |a_i|\, |\psi_i|$$

when x is common to R and u, and this proves (2). If there is an infinite number of ϕ_i, then the existence of A and U follows from the uniformity, in i, of the order relation. (3) can be deduced from (2) since under the circumstances envisaged we may take U above to be contained in U_0 and then

$$A\Sigma|a_i|\,|\psi_i| \leq A\Sigma|a_i|\,\psi = A_1\,\psi$$

where $A_1 = A\Sigma|a_i|$ is a finite constant. The proof of (4) is similar to that of (2).

Order relations may be *integrated* either with respect to the independent variable or with respect to parameters. For the sake of simplicity we shall restrict ourselves to integrals with respect to real variables. Extensions to complex and abstract variables are possible.

Let x be a real variable, let R be the interval $a < x < b$, and let $\phi = O(\psi)$ as $x \to b$. If ϕ and ψ are measurable in R then

$$(5)\quad \int_x^b \phi(t)\,dt = O\left(\int_x^b |\psi(t)|\,dt\right) \quad \text{as } x \to b.$$

Proof: If $\int_x^b |\psi(t)|\,dt = \infty$, there is nothing to prove. If $\int_x^b |\psi|\,dt < \infty$ for some x, then A and X exist so that $\int_x^b |\psi|\,dt < \infty$ and $|\phi(x)| \leq A|\psi(x)|$ for $X < x < b$, and hence

$$\left|\int_x^b \phi(t)\,dt\right| \leq \int_x^b |\phi(t)|\,dt \leq A\int_x^b |\psi(t)|\,dt \quad \text{for} \quad X < x < b.$$

Let x be a variable element of the set R in a Hausdorff space, let y be a real parameter, $a < y < \beta$, and let $\phi(x, y) = O(\psi(x, y))$, uniformly in y, as $x \to x_0$. If for each fixed x in R, ϕ and ψ are measurable functions of y in $a < y < \beta$ then

$$(6)\quad \int_\alpha^\beta \phi(x, y)\,dy = O\left(\int_\alpha^\beta |\psi(x, y)|\,dy\right) \quad \text{as} \quad x \to x_0.$$

The proof is similar to that of (5). On account of the uniformity of the O-symbol, A and U are independent of y, $|\phi| < A|\psi|$, and (6) follows by integration of this inequality with respect to y.

It is in general not permissible to *differentiate* order relations either with respect to the independent variable or with respect to parameters. However, some general results on the differentiation of order relations exist in the case of analytic functions of a complex variable (see sec. 1.6).

We conclude this section with a few formulas concerning combinations of order relations

(7) $O(O(\phi)) = O(\phi)$

(8) $O(o(\phi)) = o(O(\phi)) = o(o(\psi)) = o(\psi)$

(9) $O(\phi)\, O(\psi) = O(\phi\psi)$

(10) $O(\phi)\, o(\psi) = o(\phi)\, o(\psi) = o(\phi\psi)$

(11) $O(\phi) + O(\phi) = O(\phi) + o(\phi) = O(\phi)$

(12) $o(\phi) + o(\phi) = o(\phi)$

The proof of these formulas is immediate, and they can be extended to combinations of any finite number of order symbols.

In referring to the above rules we shall quote the number of the equation which expresses the final conclusion, and we shall use the same number to indicate the corresponding rule for the o-symbol. For instance, (1) will indicate either the rule that $\phi = O(\psi)$ and $a > 0$ imply $|\phi|^{\alpha} = O(|\psi|^{\alpha})$ or the rule that $\phi = o(\psi)$ and $a > 0$ imply $|\phi|^{\alpha} = o(|\psi|^{\alpha})$.

1.2. Asymptotic sequences

In this section R, x, x_0, ϕ have the same meaning as in sec. 1.1. A finite or infinite sequence of functions, ϕ_1, ϕ_2, ... , will be abbreviated as $\{\phi_n\}$.

The sequence of functions $\{\phi_n\}$ *is called an asymptotic sequence for* $x \to x_0$ *in* R *if for each* n, ϕ_n *is defined in* R *and* $\phi_{n+1} = o(\phi_n)$ *as* $x \to x_0$ *in* R.

It the sequence is infinite and $\phi_{n+1} = o(\phi_n)$ uniformly in n, then $\{\phi_n\}$ is said to be an *asymptotic sequence uniformly in n*. If the ϕ_n depend on parameters and $\phi_{n+1} = o(\phi_n)$ uniformly in the parameters, then $\{\phi_n\}$ is said to be an *asymptotic sequence uniformly in the parameters*.

We proceed to give some examples of asymptotic sequences in which x is a complex variable, R is the complex plane except when otherwise specified, and S_Δ is the sector defined in sec. 1.1.

(i) $\{(x - x_0)^n\}$, $x \to x_0$;

(ii) $\{x^{-n}\}$, $x \to \infty$;

(iii) $\{x^{-\lambda_n}\}$, $x \to \infty$ in S_Δ,
where $\operatorname{Re} \lambda_{n+1} > \operatorname{Re} \lambda_n$ for each n;

(iv) $\{x^{-\lambda_n}\}$, $x \to \infty$,
λ_n real and $\lambda_{n+1} > \lambda_n$ for each n;

(v) $\{e^x x^{-\lambda_n}\}$
and x and λ_n either as in (iii) or as in (iv);

(vi) $\{e^{-nx} x^{-\lambda_n}\}$, $x \to \infty$ in S_Δ,
and either $\Delta \geq 0$ and the λ_n are as in (iii) or $\Delta > 0$ and the λ_n are arbitrary:

(vii) $\{\Gamma(x)/\Gamma(x+n)\}$, $x \to \infty$ in S_Δ, $\Delta > -\pi/2$.

The reader should verify that each of the sequences (i) to (vii) is an asymptotic sequence, and should justify the restrictions imposed upon Δ and λ_n in these examples. Why is (iii), with arbitrary Im λ_n, not an asymptotic sequence for $x \to \infty$ in the complex plane (without restriction to some S_Δ)? The infinite sequence $\{\Gamma(x-n)/\Gamma(x+n)\}$, $n = 1, 2, \ldots$ is not an asymptotic sequence for $x \to \infty$ in any region including unbounded portions of the real axis, but it is an asymptotic sequence for $x \to \infty$ in any region whose closure lies entirely in the upper or the lower half-plane. The finite sequence $\{\Gamma(x-n)/\Gamma(x+n)\}$, $n = 1, 2, \ldots, N$, is an asymptotic sequence for $x \to \infty$ in any R.

From given asymptotic sequences new such sequences may be obtained by processes which are largely based on the operational rules of order symbols given in sec. 1.1. In describing some of these processes we shall restrict ourselves to real variables although extensions to more general variables are possible. In most cases x_0 and R will not be mentioned: in such cases they are fixed.

Any subsequence of an asymptotic sequence is an asymptotic sequence. The proof follows from 1.1 (8).

If $\{\phi_n\}$ *is an asymptotic sequence and* $a > 0$, *then* $\{|\phi_n|^a\}$ *is an asymptotic sequence.* The proof follows from 1.1 (1).

Two sequences, $\{\phi_n\}$ and $\{\psi_n\}$, so connected that $\phi_n = O(\psi_n)$ and $\psi_n = O(\phi_n)$ for each n, are said to be *equivalent*. *If* $\{\phi_n\}$ *and* $\{\psi_n\}$ *are equivalent sequences and* $\{\phi_n\}$ *is an asymptotic sequence, then* $\{\psi_n\}$ *is also an asymptotic sequence.* To prove that $\{\psi_n\}$ is asymptotic we remark that

$$\psi_{n+1} = O(\phi_{n+1}) = O(o(\phi_n)) = O(o(O(\psi_n))) = o(\psi_n)$$

by 1.1 (8).

If $\{\phi_n\}$ *and* $\{\psi_n\}$ *are asymptotic sequences containing the same number of functions, then* $\{\phi_n \psi_n\}$ *is an asymptotic sequence.* The proof follows from 1.1 (10).

If $\{\phi_n\}$, $n = 1, \dots, N$, *is an asymptotic sequence,* $a_{n,i}$, $n = 1, \dots, N$; $i = 0, 1, \dots, k < N$ *is a set of positive constants,* $a_{n+1,i} \leq a_{n,i}$ *for all* n, i *and*

$$(1) \quad \psi_n = \sum_{i=0}^{k} a_{n,i} |\phi_{n+i}| \qquad n = 1, \dots, N - k,$$

then $\{\psi_n\}$ *is an asymptotic sequence.*

In this statement N may be finite or infinite; k is finite. To prove the statement we remark that it follows from k being finite that for any n and any $\epsilon > 0$ there exists a neighborhood U_ϵ of x_0 so that $|\phi_{r+1}| \leq \epsilon|\phi_r|$ in the common part of U_ϵ and R for $r = n, n+1, \dots, n+k$. We then have

$$(2) \quad \psi_{n+1} = \sum_{i=0}^{k} a_{n+1,i} |\phi_{n+i+1}| \leq \epsilon \sum_{i=0}^{k} a_{n,i} |\phi_{n+i}| = \epsilon\, \psi_n$$

The extension to infinite sums is contained in the following theorem.

Let $\{\phi_n\}$ *be an asymptotic sequence uniformly in* n, *let* $a_{n,i}$, $n = 1, 2, \dots$, $i = 0, 1, \dots$ *be a set of positive constants such that* $a_{n+1,i} \leq a_{n,i}$ *for all* n, i *and put*

$$(3) \quad \psi_n = \sum_{i=0}^{\infty} a_{n,i} |\phi_{n+i}| \qquad n = 1, 2, \dots$$

If the infinite series for ψ_1 *converges in some neighborhood of* x_0, *then there is a subset* R_0 *of* R *so that* x_0 *is a limit point of* R_0, *all infinite series* (3) *converge in* R_0, *and* $\{\psi_n\}$ *is an asymptotic sequence for* $x \to x_0$ *in* R_0, *uniformly in* n.

Proof: From the uniform asymptotic property of $\{\phi_n\}$, it follows that there exists a subset R_1 of R so that x_0 is a limit point of R_1 and $|\phi_{n+1}| \leq |\phi_n|$ for all x in R_1 and all n. For x in R_1

$$\sum a_{n+1,i} |\phi_{n+1+i}| \leq \sum a_{n,i} |\phi_{n+i}| < \cdots < \sum a_{1,i} |\phi_{i+1}|$$

so that all infinite series (3) are dominated by the series for ψ_1. If the series for ψ_1 converges in a subset R_2 of R, and x_0 is a limit point of R_2, we take R_0 to be the common part of R_1 and R_2. All functions ψ_n are defined in R_0, x_0 is a limit point of R_0, and on account of the uniform asymptotic property of $\{\phi_n\}$, equation (2) with $k = \infty$ holds uniformly in n.

New asymptotic sequences may be formed by integration in two different ways.

If $\{\phi_n(x, y)\}$ *is an asymptotic sequence uniformly in* y, $\alpha < y < \beta$, *for* $x \to x_0$ *in* R, *and if all integrals*

$$(4) \quad \Phi_n(x) = \int_\alpha^\beta |\phi_n(x, y)|\, dy$$

exist, then $\{\Phi_n\}$ *is an asymptotic sequence.* The proof follows from 1.1 (6). As in the case of (3), it is sufficient to assume that all $\phi_n(x, y)$ are measurable functions of y, and that ϕ_1 is integrable. The integrability of all ϕ_n, possibly for a more restricted set R_0, then follows by showing that $\int|\phi_n|dy$ is dominated by $\int|\phi_1|\, dy$.

If x *is a real variable,* R *is the interval* $a < x < b$, $\{\phi_n\}$ *is an asymptotic sequence for* $x \to b$, *and if all integrals*

$$(5) \quad \Phi_n(x) = \int_x^b |\phi_n(t)|\, dt$$

exist, then $\{\Phi_n\}$ *is an asymptotic sequence for* $x \to b$. The proof follows from 1.1 (5), and it is again sufficient to assume that all ϕ_n are measurable and ϕ_1 is integrable; the result then follows at least for some interval $a_1 < x < b$.

Note that the differentiation of an asymptotic sequence does not necessarily yield an asymptotic sequence. For instance, take

$$\phi_n = x^{-n}[a + \cos(x^n)] \qquad n = 1, 2, \ldots .$$

Then $\{\phi_n\}$ is an asymptotic sequence for $x \to \infty$ on the real axis, but $\{\phi_n'\}$ is not an asymptotic sequence.

1.3. Asymptotic expansions

In this section and in the following sections, x, x_0, R have the same meaning as in sec. 1.1; $\{\phi_n\}$, $\{\psi_n\}$, $\{\chi_n\}$, ... , are always asymptotic sequences for $x \to x_0$ in R; $f(x)$, $g(x)$, $h(x)$, ... are numerical functions of x defined in R; and a, b, c, ... are constants (i.e., independent of x).

The (formal) series $\sum a_n \phi_n(x)$ *is said to be an asymptotic expansion to* N *terms of* $f(x)$ *as* $x \to x_0$ *if*

$$(1) \quad f(x) = \sum_{n=1}^{N} a_n \phi_n(x) + o(\phi_N) \quad \text{as} \quad x \to x_0 .$$

An asymptotic expansion to N terms will often be indicated as

(2) $f(x) \sim \Sigma a_n \phi_n(x)$ to N terms as $x \to x_0$ in R,

and the qualifying phrase "in R" will often be omitted. An asymptotic expansion to 1 term will be written as

(3) $f(x) \sim a_1 \phi_1(x) \qquad x \to x_0$

and will be called an *asymptotic representation*; and an asymptotic expansion to any number of terms (i.e., with $N = \infty$) will be written

(4) $f(x) \sim \Sigma a_n \phi_n(x) \qquad x \to x_0$

and called an *asymptotic expansion*. An asymptotic expansion may be convergent or divergent. In most textbooks only the cases $N = 1$ and $N = \infty$ are discussed, but we shall let N stand for any positive integer.

If an asymptotic expansion to N terms, with N finite, involves certain parameters, we shall say that it holds *uniformly* in these parameters if the remainder in (1) in $o(\phi_N)$ uniformly in the parameters. An asymptotic expansion ($N = \infty$) involving certain parameters will be said to hold uniformly in these parameters if $f - \sum_{n=1}^{M} a_n \phi_n = o(\phi_M)$ uniformly in the parameters for each sufficiently large M (but not necessarily uniformly in M.).

The formal (finite, or infinite) series $\Sigma a_n \phi_n$ will be called an *asymptotic series*. If $\phi_n = x^{-\lambda_n}$, we shall speak of an asymptotic *series of powers*, and if $\phi_n = x^{\pm n}$, of an asymptotic *power series*. For instance, $\Sigma (n-1)! (-x)^{n-1}$ is the asymptotic power series expansion for $x \to 0$ in $S_{-\pi/2+\epsilon}$ of the function $f(x)$ defined by equation (3) of the Introduction. Some authors speak of asymptotic power series when $\phi_n = \phi_0(x)\, x^{\pm n}$ but it is more appropriate to call the series divided by $\phi_0(x)$ an asymptotic power series.

From (1) it follows that the coefficients in an asymptotic expansion to N terms may be computed by means of the recurrence formula

(5) $a_m = \lim_{x \to x_0} \{[f(x) - \sum_{n=1}^{m-1} a_n \phi_n(x)]/\phi_m(x)\} \qquad m = 1, \ldots, N$

Conversely,* suppose that we have $N+1$ functions,

$$f(x),\ \phi_1(x),\ \ldots,\ \phi_N(x)$$

defined in R. If (5) holds and $a_m \neq 0$ *for* $m = 1, \ldots, N$ *then* $\{\phi_n\}$ *is an asymptotic sequence for* $x \to x_0$, and $\sum a_n \phi_n$ *is an asymptotic expansion to N terms of* $f(x)$ *as* $x \to x_0$.

To prove that $\{\phi_n\}$ is an asymptotic sequence we have to show that $\phi_{m+1} = o(\phi_m)$ for $m = 1, \ldots, N-1$. Now from (5),

$$f - \sum_{n=1}^{m} a_n \phi_n = o(\phi_m),$$

and if we replace m by $m+1$ in (5) we have

$$f - \sum_{n=1}^{m} a_n \phi_n = a_{m+1} \phi_{m+1} + o(\phi_{m+1}).$$

Comparing the last two equations we find

$$[a_{m+1} + o(1)]\, \phi_{m+1} = o(\phi_m).$$

If $a_{m+1} \neq 0$ then $a_{m+1} + o(1) \neq 0$ for x in some neighborhood of x_0, and we may divide by this factor to see that $\phi_{m+1} = o(\phi_m)$. Thus $\{\phi_n\}$ is an asymptotic sequence. Moreover, (5) with $m = N$ shows that (1) holds, and $\sum a_n \phi_n$ is an asymptotic expansion to N terms of f.

If $\sum a_n \phi_n(x)$ is an asymptotic expansion to N terms of $f(x)$, then the same formal series will also provide an asymptotic expansion to any lesser number of terms of the same function. We also have the somewhat sharper result

$$(6)\quad f(x) = \sum_{n=1}^{M} a_n \phi_n(x) + O(\phi_{M+1}) \qquad x \to x_0, \quad M = 1, \ldots, N-1$$

which is an immediate consequence of (1).

With x_0 and R fixed, (5) shows that the asymptotic expansion to a given number of terms of a given function is *unique* if the asymptotic sequence is given. On the other hand, one and the same function may have asymptotic expansions involving two different asymptotic sequences, and the two sequences need not be equivalent in the sense of

* This theorem was suggested by Dr. A.G. Mackie.

sec. 1.2. For instance,

$$\frac{1}{1+x} \sim \Sigma\,(-1)^{n-1}\,x^{-n} \qquad x \to \infty$$

$$\frac{1}{1+x} \sim \Sigma\,(x-1)\,x^{-2n} \qquad x \to \infty$$

$$\frac{1}{1+x} \sim \Sigma(-1)^{n-1}\,(x^2 - x + 1)\,x^{-3n} \qquad x \to \infty$$

In this example all three asymptotic expansions are convergent series when $|x| > 1$. It often happens that some asymptotic expansions of a function diverge while others converge. The transformation of divergent asymptotic expansions into convergent ones is of great analytical, although of very little computational interest. Transformations of asymptotic expansions into convergent expansions or else into expansions more suited to numerical computation have been investigated among others by Airey (1937), van der Corput (1951), Miller (1952), van Wijngaarden (1953), Watson (1912b).

An asymptotic expansion does not determine its "sum", $f(x)$, uniquely. For instance, the functions $(1+x)^{-1}$, $(1+e^{-x})/(1+x)$, $(1+e^{-\sqrt{x}}+x)^{-1}$ all possess the asymptotic expansion $\Sigma(-1)^{n-1}\,x^{-n}$ as $x \to \infty$ in S_Δ, $\Delta > 0$. A given (finite or infinite) asymptotic sequence, $\{\phi_n\}$, for $x \to x_0$ in R establishes an *equivalence relation* among functions defined in R: $f(x)$ and $g(x)$ are *asymptotically equal* with respect to $\{\phi_n\}$ if

$$f(x) - g(x) = o(\phi_n) \quad \text{as} \quad x \to x_0 \quad \text{in} \quad R,$$

for all n occurring in the sequence. An asymptotic series represents a class of asymptotically equal functions rather than a single function.

1.4. Linear operations with asymptotic expansions

If $f \sim \Sigma\, a_n\, \phi_n$ *and* $g \sim \Sigma\, b_n\, \phi_n$, *both to* N *terms, and if* α, β *are constants, then*

$$(1) \quad \alpha f(x) + \beta g(x) \sim \Sigma\,(\alpha a_n + \beta b_n)\,\phi_n(x) \quad \text{to } N \text{ terms.}$$

The proof of this theorem is obvious, as is its extension to a linear combination of an arbitrary finite number of asymptotic expansions. The extension to an infinite series of asymptotic expansions is as follows:

If $f_i(x) \sim \Sigma\, a_{n,\,i}\,\phi_n(x)$ *to* N *terms, uniformly in* i, $i = 1, 2, \dots$, *and if the* a_i *are constants for which* $\Sigma\, a_i$ *converges absolutely and*

$$(2) \quad A_n = \sum_{i=1}^{\infty} a_{n,\,i}\, a_i$$

converges for each n, *then* $\Sigma\, a_i f_i(x)$ *converges in some neighborhood of* x_0, *and*

$$(3) \quad F(x) = \sum_{i=1}^{\infty} a_i f_i(x) \sim \Sigma\, A_n\, \phi_n(x) \quad \text{to } N \text{ terms.}$$

Proof: We have

$$f_i - \sum_{n=1}^{N} a_{n,\,i}\,\phi_n = o(\phi_N)$$

uniformly in i, and $\Sigma\, |a_i| < \infty$. By 1.1 (3),

$$\sum_{i=1}^{\infty} a_i(f_i - \sum_{n=1}^{N} a_{n,\,i}\,\phi_n) = o(\phi_N),$$

and the infinite series on the left is convergent at least in some neighborhood of x_0. Adding $\sum_{n=1}^{N} A_n\, \phi_n$ to both sides we have (3) when $N < \infty$. If $N = \infty$ then both the assumptions and the conclusion hold for all sufficiently large N, and hence $\Sigma\, A_n\, \phi_n$ is an asymptotic expansion of $F(x)$.

More generally, we may extend (1) to finite or infinite asymptotic expansions.

Let $\{\phi_n\}$, $n = 1, \dots, N < \infty$ *and* $\{\psi_m\}$, $m = 1, \dots, M \leq \infty$ *be asymptotic sequences for the same* R, x_0; *and let* $\phi_N = O(\psi_m)$ *for each* m: *if* $f \sim \Sigma\, a_n\, \phi_n$ *to* N *terms and for each* n, $\phi_n \sim \Sigma\, b_{mn}\, \psi_m$ *to* M *terms, then*

$$(4) \quad f(x) \sim \Sigma\, c_m\, \psi_m(x) \quad \text{to } M \text{ terms,}$$

where

$$(5) \quad c_m = \sum_{n=1}^{N} a_n\, b_{mn}\ .$$

Let $\{\phi_n\}$, $n = 1, 2, \ldots$, *and* $\{\psi_m\}$, $m = 1, \ldots , M \leq \infty$ *be asymptotic sequences; suppose that for each n there is an integer* $\mu(n) \leq M$ *so that* $\mu(n) \to M$ *as* $n \to \infty$ *and* $\phi_n = O(\psi_{\mu(n)})$: *if* $f \sim \Sigma\, a_n \phi_n$, $\phi_n \sim \Sigma\, b_{mn} \psi_m$ *to M terms uniformly in n,* $\Sigma\, a_n$ *is absolutely convergent, and the infinite series in* (5) *is convergent for each m; then* (4) *holds.*

The proof for M, N finite is immediate, since then

$$f(x) = \sum_{n=1}^{N} a_n \phi_n + o(\phi_N)$$

$$= \sum_{n=1}^{N} a_n \left(\sum_{m=1}^{M} b_{mn} \psi_m + o(\psi_M)\right) + o(\psi_M)$$

$$= \sum_{m=1}^{M} c_m \psi_m + o(\psi_M)$$

by 1.1 (3). If $M = \infty$, then the same reasoning holds for any M, and hence $\Sigma\, c_m \psi_m$ is an asymptotic expansion (with $M = \infty$) of $f(x)$. In the extension to $N = \infty$ we use the extension of 1.1 (3) for infinite series.

We now turn to the *integration* of asymptotic expansions either with respect to a real parameter y, or with respect to the variable x. In the latter case x will be assumed to be a real variable.

If $f(x, y) \sim \Sigma\, a_n(y)\, \phi_n(x)$ *to N terms, uniformly in y,* $\alpha < y < \beta$, *if* $f(x, y)$, *for each fixed x, and* $a_n(y)$, *for each fixed n, is a measurable function of y, and if* $h(y)$ *is an integrable function of y for which each of the integrals*

$$(6) \quad A_n = \int_{\alpha}^{\beta} h(y)\, a_n(y)\, dy$$

exists, then also the integral

$$(7) \quad F(x) = \int_{\alpha}^{\beta} h(y)\, f(x, y)\, dy$$

exists for each x in some neighborhood of x_0, *and*

$$(8) \quad F(x) \sim \Sigma\, A_n\, \phi_n(x) \quad \text{to } N \text{ terms.}$$

The proof is very similar to that given above for infinite series, except that 1.1 (6) must be used instead of 1.1 (3). Some generalizations of this theorem are obvious: the interval (α, β) may be replaced by any measurable set, of finite or infinite measure, and there is a similar result for multiple integrals.

Let x be a real variable, let R be the interval $a < x < b$, let $\{\phi_n(x)\}$ be an asymptotic sequence of positive functions for $x \to b$, and assume that each of the integrals

(9) $\Phi_n(x) = \int_x^b \phi_n(t)\, dt$

exists. If $f(x) \sim \Sigma a_n \phi_n(x)$ to N terms as $x \to b$, and $f(x$ is a measurable function, then

(10) $F(x) = \int_x^b f(t)\, dt$

exists in some interval $c < x < b$, and

(11) $F(x) \sim \Sigma a_n \Phi_n(x)$ to N terms as $x \to b$.

The proof follows from 1.1(5).

It is, in general, not permissible to differentiate asymptotic expansions either with respect to the variable x, or with respect to parameters. Some general results on the differentiation of asymptotic expansions of analytic functions of a complex variable exist and will be given in sec. 1.6.

1.5. Other operations with asymptotic expansions

Multiplication of asymptotic series does not in general lead to an asymptotic series, for in the formal product of $\Sigma a_n \phi_n$ and $\Sigma b_n \phi_n$ all products $\phi_m \phi_n$ occur, and it is in general not possible to arrange the system of functions $\{\phi_m \phi_n\}$, $m, n = 1, \dots, N$ so as to obtain an asymptotic sequence. There are, however, important special asymptotic sequences $\{\phi_n\}$ with the property that the products $\phi_m \phi_n$ either form an asymptotic sequence, or else possess asymptotic expansions in terms of an asymptotic sequence (which need not be $\{\phi_n\}$). First we shall prove a general result on the multiplication of two asymptotic expansions.

Let $\{\phi_n\}$, $n = 1, \dots, N$, $\{\psi_m\}$, $m = 1, \dots, M$, and $\{\chi_k\}$, $k = 1, \dots, K$ be three asymptotic sequences such that $\phi_1 \psi_M = O(\chi_K)$, $\phi_N \psi_1 = O(\chi_K)$, and

(1) $\phi_n \psi_m \sim \Sigma c_{nmk} \chi_k$ to K terms.

If $f \sim \sum a_n \phi_n$ *to N terms and* $g \sim \sum b_m \psi_m$ *to M terms, then* $fg \sim \sum C_k \chi_k$ *to K terms, where*

$$(2) \quad C_k = \sum_{n=1}^{N} \sum_{m=1}^{M} a_n b_m c_{nmk} .$$

Here K may be finite or infinite; N, M are finite. The result remains true if M, or N, or both M and N are infinite, provided that each of the infinite series (or double series) in (2) *converges.*

The coefficients C_k are those obtained upon multiplication of $\sum a_n \phi_n$ and $\sum b_m \psi_m$ and substitution of (1) so that instead of (2) we may say "where the coefficients C_k are obtained by formal substitution", and this description will be used in similar cases throughout this section.

We first prove the theorem for finite N, M, K.

$$fg = [\sum_{n=1}^{N} a_n \phi_n + o(\phi_N)][\sum_{m=1}^{M} b_m \psi_m + o(\psi_M)]$$

$$= \sum_{n=1}^{N} \sum_{m=1}^{M} a_n b_m \phi_n \psi_m + o(\phi_1 \psi_M) + o(\phi_N \psi_1)$$

$$= \sum_{k=1}^{K} C_k \chi_k + o(\chi_K) + o(\phi_1 \psi_M) + o(\phi_N \psi_1)$$

by (1) and (2). This proves the result. If (1) holds to any number of terms and $\phi_1 \psi_M$ and $\phi_N \psi_1$ are $O(\chi_k)$, for any k, then the above computation holds for any K, and the extension to $K = \infty$ holds. The extension to infinite M, N can similarly be justified provided that the infinite series defining C_k converge.

A sequence of functions, $\{\phi_n\}$, $n = 1, \dots, N$ *will be called a multiplicative asymptotic sequence if* $\{\phi_n\}$ *is an asymptotic sequence,* $\phi_1 = O(1)$ *and* $\phi_n \phi_m \sim \sum c_{nmk} \phi_k$ *to N terms,* $m, n = 1, \dots, N$. In the case of a multiplicative asymptotic sequence the former result on the multiplication of asymptotic expansions can be extended considerably.

If $\{\phi_n\}$, $n = 1, \dots, N$ *is a multiplicative asymptotic sequence,*

$$f_i \sim \sum a_{n,i} \phi_n \quad \text{to } N \text{ terms} \qquad i = 1, \dots, k,$$

and $P(z_1, \dots, z_k)$ *is a polynomial in the k complex variables* $z_1, \dots, z_k$, *then* $F(x) = P(f_1, \dots, f_k)$ *possesses an asymptotic expansion* $\sum A_n \phi_n$ *to N terms, and the coefficients* A_n *may be computed by formal substitution.*

To prove this theorem we remark that in the case of a multiplicative asymptotic sequence $\{\phi_n\}$ we have $\phi_1\ \phi_N = O(\phi_N)$, and we also have an asymptotic expansion of $\phi_n\ \phi_m$ to N terms. By our general theorem, it follows from $f \sim \Sigma\ a_n\ \phi_n$ to N terms and $g \sim \Sigma\ b_n\ \phi_n$ to N terms that fg possesses an asymptotic expansion $\Sigma\ c_n\ \phi_n$ to N terms, and the coefficients c_n may be computed by formal substitution. The evaluation of any polynomial $P(f_1, \ldots, f_k)$ can be reduced to a finite number of operations each of which involves either a linear combination, or the multiplication, of two asymptotic expansions. Each of these operations preserves the asymptotic character of the expansion, and in each operation the resulting expansion may be computed by formal substitution. Hence the theorem.

The result obtained for polynomials can, under certain circumstances, be extended to (convergent) power series, and even to asymptotic power series. For the sake of simplicity, we shall restrict ourselves to the case of a single variable z; there is a generalization to the case of several variables.

Let $\{\phi_n\}$, $n = 1, \ldots, N$ be a multiplicative asymptotic sequence such that $\phi_1 = o(1)$, and $|\phi_1|^M = O(\phi_N)$ for some positive integer M. *If $f(z) \sim \Sigma\ c_m\ z^m$ to M terms as $z \to 0$ in the complex plane, and*

$$z = z(x) \sim \Sigma\ a_n\ \phi_n \quad \textit{to } N \textit{ terms as } x \to x_0 \textit{ in } R,$$

then $F(x) = f(z(x))$ possesses an asymptotic expansion $\Sigma\ A_n\ \phi_n$ to N terms as $x \to x_0$, and the coefficients A_n may be computed by formal substitution.

Proof: From the assumptions it follows that z^m possesses an asymptotic expansion $\Sigma\ b_{mn}\ \phi_n$ to N terms, and also that $z^M = O(\phi_N)$. Hence we can apply the theorem in sec. 1.4 on the substitution of an asymptotic expansion into an asymptotic expansion.

An important particular case concerns functions $f(x)$ which possess asymptotic expansions of the form

$$(3) \quad f(x) = c + \sum_{n=1}^{N} a_n\ \phi_n + o(\phi_N).$$

The theorem shows that $[f(x)]^{-1}$ also possesses an asymptotic expansion of this form provided that $c \neq 0$ and $\{\phi_n\}$ satisfies the assumptions of the theorem. In other words, asymptotic expansions of the form (3) may be divided. This enables us to extend the last theorem but one from polynomials to rational functions.

If $\{\phi_n\}$, $n = 1, \ldots, N$ *is a multiplicative asymptotic sequence,* $\phi_1 = o(1)$, $|\phi_1|^M = O(\phi_N)$ *for some M,* $f_i \sim \Sigma\, a_{n,i}\, \phi_n$ *to N terms,* $i = 1, \ldots, k$, *and* $P(z_1, \ldots, z_k)$ *is a rational function in the k complex variables* $z_1, \ldots, z_k$ *such that the denominator is different from zero when* $z_1 = z_2 = \cdots = z_k = 0$; *then* $F(x) = P(f_1, \ldots, f_k)$ *possesses an asymptotic expansion* $A_0 + \Sigma\, A_n \phi_n$ *to N terms, and the coefficients* A_n *may be computed by formal substitution.*

Under the same conditions we also have an asymptotic expansion for $g(F(x))$ if $g(\zeta)$ is a function of the complex variable ζ which is regular in some neighborhood of $\zeta_0 = P(0, \ldots, 0)$. In this manner we may justify the asymptotic expansions of expressions such as $\exp[P(f_1, \ldots, f_k)]$.

1.6 Asymptotic power series

The sequence of functions $\{x^{-n}\}$, $n = 0, 1, 2, \ldots$ or $n = 1, 2, \ldots$ is a multiplicative asymptotic sequence for $x \to \infty$ in any region of the complex plane which does not include the origin. This sequence satisfies all the conditions imposed upon asymptotic sequences in the two preceding sections, except that in some of the theorems of sec. 1.5, $n = 0$ must be excluded. Besides, this system has some special properties.

The asymptotic expansion

$$(1)\quad f(x) \sim a_0 + \frac{a_1}{x} + \frac{a_2}{x^2} + \cdots \quad \text{to } N \text{ terms as } x \to \infty$$

is an *asymptotic power series*. From the results of sections 1.5 and 1.6 it follows that an asymptotic power series expansion may be multiplied by a constant, and that two such expansions may be added or multiplied, and also divided provided that $a_0 \neq 0$ in the expansion in the denominator. Asymptotic power series may be substituted in finite linear combinations, in polynomials, in rational functions provided that the denominator does not vanish as $x \to \infty$, and in asymptotic or convergent power series $\Sigma\, c_n z^n$, $z \to 0$, provided that in the expansion (1) of $z = f(x)$ we have $a_0 = 0$. Substitution of (1) in other types of convergent or asymptotic series is valid under the conditions set out in sec. 1.5. In all these cases the coefficients of the new expansion are obtained by formal substitution and a rearrangement of terms. An asymptotic power series expansion (1) which is valid uniformly in a parameter may be integrated with respect to this parameter. Lastly, *if* (1) *holds, then* $f(x) - a_0 - a_1/x$ *is integrable, and*

$$(2) \quad F(x) = \int_x^\infty \left[f(t) - a_0 - \frac{a_1}{t} \right] dt$$

$$\sim \frac{a_2}{x} + \frac{a_3}{2x^2} + \frac{a_4}{3x^3} + \cdots \quad \text{to } N-2 \text{ terms as } x \to \infty.$$

A simple corollary of this last result is the following theorem on differentiation. *If $f(x)$ in* (1) *is differentiable and if $f'(x)$ possesses an asymptotic power series expansion, then*

$$(3) \quad f'(x) \sim -\frac{a_1}{x^2} - \frac{2a_2}{x^3} - \frac{3a_3}{x^4} - \cdots, \quad \text{to } N-1 \text{ terms as } x \to \infty.$$

In the case of *analytic functions* a more definite statement can be made in that it is not necessary to assume that $f'(x)$ possesses an asymptotic power series expansion. Let R be the region

$$|x| > a, \qquad \alpha < \arg x < \beta,$$

let $a_1 > a$, $\alpha < \alpha_1 < \beta_1 < \beta$, and let R_1 be the region

$$|x| > a_1, \qquad \alpha_1 < \arg x < \beta_1.$$

If $f(x)$ is regular in R and (1) *holds uniformly in* $\arg x$ *as $x \to \infty$ in R, then* (3) *holds uniformly in* $\arg x$ *as $x \to \infty$ in R_1.* The proof of this theorem follows from Cauchy's integral formula for the derivative,

$$(4) \quad f'(x) = \frac{1}{2\pi i} \int_C \frac{f(z)}{(x-z)^2}\, dz.$$

For given R, R_1, there exists an $\epsilon > 0$ so that for each x in R_1, the circle with center x and radius $\epsilon|x|$ is in R, and we may take this circle as the contour of integration in (4). Along the circle, $z = x + \epsilon x e^{it}$, and $0 \le t \le 2\pi$, so that (4) becomes

$$(5) \quad f'(x) = \frac{1}{2\pi x \epsilon} \int_0^{2\pi} e^{-it} f[x(1+\epsilon e^{it})]\, dt.$$

Now, $e^{-it} f[(x + \epsilon e^{it})]$ possesses an asymptotic power series expansion uniformly in t, and this may be integrated with respect to t, showing that $f'(x)$ possesses an asymptotic power series expansion which turns out to be (3).

Asymptotic power series expansions are usually valid in sectorial regions, and analytic functions possess different asymptotic expansions in different sectors (*Stokes' phenomenon*). That something like this must happen, except in the case of an analytic function which is regular at infinity, follows from the following theorem.

If $f(x)$ *is single-valued and regular when* $|x| > a$, *and* (1) *holds for all values of* arg x, *then the power series in* (1) *converges for sufficiently large values of* $|x|$, *and its sum if* $f(x)$. To prove this, we set $x = 1/\xi$ and $g(0) = a_0$, $g(\xi) = f(1/\xi)$, $0 < |\xi| < |a|^{-1}$. Then $g(\xi)$ is a single-valued continuous function in $|\xi| < |a|^{-1}$, and is regular except possibly at $\xi = 0$. However, at $\xi = 0$, g has certainly no pole, nor an essential singularity, since it is bounded in any neighborhood of $\xi = 0$. Thus, $g(\xi)$ is regular at $\xi = 0$ and possesses a Maclaurin expansion. From the uniqueness theorem on asymptotic expansions it follows that (1), with $x = 1/\xi$, must be the Maclaurin expansion.

1.7. Summation of asymptotic series

It has been pointed out in sec. 1.3 that an asymptotic sequence $\{\phi_n\}$ determines an equivalence relation between functions defined in R. Two functions defined in R are asymptotically equal if their difference is $o(\phi_n)$ for all n. Asymptotically equal functions possess identical asymptotic expansions, and given an asymptotic expansion $f \sim \Sigma\, a_n\, \phi_n$, we may define the *class of all functions which are asymptotically equal to f* as the *sum of the asymptotic series* $\Sigma\, a_n\, \phi_n$.

We shall conclude this chapter by proving that *every asymptotic series possesses a sum.* Results of this nature have been proved for asymptotic power series by Borel and Carleman (1926), for series dominated by an asymptotic series of powers by van der Corput (1954b), and for asymptotic series of analytic functions by Carleman (1926). The proof given below is an adaptation of van der Corput's proof.

An asymptotic series is a formal finite or infinite series $\Sigma\, a_n\, \phi_n(x)$ where $\{\phi_n\}$ is an asymptotic sequence and the a_n are constants. Since any subsequence of an asymptotic sequence is also such a sequence, we may assume that $a_n \neq 0$ for each n. The asymptotic sum of $\Sigma\, a_n\, \phi_n$ is a class of asymptotically equal functions, and we shall demonstrate the existence of the asymptotic sum by constructing a member of this class. If $\Sigma\, a_n\, \phi_n$ is a finite asymptotic series, then the sum

$$a_1\phi_1 + \cdots + a_N\phi_N$$

in the ordinary sense may be taken as a representative of the asymptotic sum. It is sufficient, then, to give the proof for an infinite asymptotic series $\Sigma\, a_n\phi_n$ in which $a_n \neq 0$ for each n.

Let U_0 be a neighborhood of x_0, and for each $n = 1, 2, \ldots$ let U_n be a neighborhood of x_0 such that the closure of U_n is in U_{n-1} and

$$|a_{n+1}\phi_{n+1}| \leq \frac{1}{2}|a_n\phi_n|$$

for all x common to U_n and R: such a neighborhood exists since

$$a_{n+1}\phi_{n+1} = o(a_n\phi_n).$$

For each n let $\mu_n(x)$ be a continuous function of x such that $0 \leq \mu_n(x) \leq 1$ in R, $\mu_n(x) = 0$ when x is outside U_n and $\mu_n(x) = 1$ when x is in U_{n+1}: such a function exists since the closure of U_{n+1} is contained in U_n. Then

$$(1) \quad |a_{n+p}\mu_{n+p}(x)\phi_{n+p}(x)| \leq 2^{-p}|a_n\phi_n(x)|$$

when x is in U_n, for this inequality holds by the construction of the U's if x is in U_{n+p}, and the left-hand side vanishes when x is outside U_{n+p}. Let

$$(2) \quad f(x) = \sum_{n=1}^{\infty} a_n\mu_n(x)\phi_n(x).$$

The series converges for all x by (1), and defines a function $f(x)$ in R. (Actually, the series terminates except for those x which are in all the U_n.) To show that $f \sim \Sigma\, a_n\phi_n$ as $x \to x_0$, fix N and let x be in the common part of U_{N+1} and R. Then $\mu_n(x) = 1$ for $n = 1, \ldots, N$, and by (1)

$$|f - \sum_{n=1}^{N} a_n\phi_n| \leq \sum_{N+1}^{\infty} |a_n\mu_n\phi_n| \leq |a_{N+1}\phi_{N+1}| \sum_{N+1}^{\infty} 2^{N+1-n}$$

$$= 2|a_{N+1}\phi_{N+1}| = o(\phi_N).$$

Thus it is seen that $\Sigma\, a_n\phi_n$ is an asymptotic expansion to any number of terms of f defined by (2). The asymptotic sum of $\Sigma\, a_n\phi_n$ is the class of all functions asymptotically equal to f.

The U_n may be constructed in such a manner that x_0 is the only point common to all the U_n in which case the series in (2) terminates for all $x \neq x_0$. If all the ϕ_n are continuous in R, also f will be continuous in R. If x is a real variable, or a point in n-dimensional Euclidean space, the $\mu_n(x)$ may be chosen as infinitely differentiable functions, and if all the ϕ_n are k times continuously differentiable ($k \leq \infty$), then $f(x)$ will also be k times continuously differentiable. Carleman has proved that for certain analytic functions ϕ_n of a complex variable x, the asymptotic sum contains a function which is an analytic function of x.

In general there is no way of ascribing a unique asymptotic sum to an asymptotic series, but under rather special circumstances it may happen that under more precise assumptions on the coefficients of the asymptotic series, and under certain restrictions on the functions $f(x)$, a unique sum may be obtained; and frequently in such cases the asymptotic series, though divergent, is in some sense *summable* to its asymptotic sum. Such theorems for asymptotic power series summed by analytic functions regular in some sectorial region were obtained by Watson (1912a) and Nevanlinna (1916).

REFERENCES

Airey, J.R., 1937: *Philos. Mag.* (7) 24, 521-552.

Borel, Émile, 1895: *Ann. Sci. École Norm. Sup.* (3) 12, 9-55.

Borel, Émile, 1899: *Ann. Sci. École Norm. Sup.* (3) 16, 8-136.

Borel, Émile, 1928: *Leçons sur les séries divergentes*, second ed. Paris.

Bromwich, T.J. I'A., 1926: *Infinite series*, second ed., McMillan, especially sec. 113ff.

Carleman, T.G.T., 1926: *Les fonctions quasi-analytiques.* Paris, especially Chap. V.

van der Corput, J.G., 1951: *Asymptotic expansions*, Parts I and II. National Bureau of Standards (Working Paper).

van der Corput, J.G., 1952: *Asymptotic expansions*, Part III. National Bureau of Standards (Working Paper).

van der Corput, J.G., 1954a: *Nederl. Akad. Wetensch., Amsterdam, Proc.* 57, 206-217.

van der Corput, J.G., 1954b: *Asymptotic Expansions* I. *Fundamental theorems of Asymptotics.* Department of Mathematics, University of California, Berkeley.

Knopp, Konrad, 1928: *Theory and application of infinite series*, especially Chap. XIV.

Miller, J.C.P., 1952: *Proc. Cambridge Philos. Soc.* 48, 243-254.

Nevanlinna, F.E.H., 1916: *Ann. Acad. Sci. Fennicae* (*A*) 12, no. 3, 81 pp.

Watson, G.N., 1912a: *Philos. Trans. Royal Soc. A*, 211, 279-313.

Watson, G.N., 1912b: *Rend. Circ. Mat. Palermo* 34, 41-88.

van Wijngaarden, A., 1953: *Nederl. Akad. Wetensch., Amsterdam, Proc.*, 56, 522-543.

CHAPTER II

INTEGRALS

There are several methods for obtaining asymptotic expansions of functions defined by definite integrals. Copson (1946) gives a survey of these; and further material is contained in van der Corput's Lectures and in the references given at the end of this chapter.

2.1. Integration by parts

Asymptotic expansions may frequently be obtained by repeated integrations by parts. As an example, let us consider the function $f(x)$ defined for $-\pi < \arg x < \pi$ by the integral

$$(1) \quad f(x) = \int_0^\infty \frac{e^{-t}}{1+xt}\, dt.$$

Integrating by parts repeatedly,

$$(2) \quad f(x) = 1 - x \int_0^\infty \frac{e^{-t}}{(1+xt)^2}\, dt$$

$$= 1 - x + 2x^2 \int_0^\infty \frac{e^{-t}}{(1+xt)^3}\, dt$$

$$= \cdots$$

$$= \sum_{n=0}^{m} (-1)^n n!\, x^n + (-1)^{m+1} (m+1)!\, x^{m+1} \int_0^\infty \frac{e^{-t}}{(1+xt)^{m+2}}\, dt$$

The last integral may be proved to be $O(1)$ as $x \to 0$ in $S_\Delta, \Delta > -\pi/2$, so that we have obtained a new derivation of Euler's asymptotic expansion discussed in the Introduction.

The field of application of this method is somewhat limited, and it is not at all easy to formulate precise theorems of sufficient generality. In what follows we shall describe some results which seem to be basic.

For any function $f(t)$, let f_m denote the m-th derivative, and f_{-m} the m-th repeated integral, so that

$$(3) \quad f_0 = f, \quad f_m = \frac{d^m f}{dt^m} \qquad m = 1, 2, \ldots$$

$$(4) \quad \frac{df_{-m}}{dt} = f_{-m+1} \qquad m = 1, 2, \ldots .$$

Note that f_{-m} contains m constants (one from each integration) which we suppose to have been chosen in some suitable manner. The formula

$$(5) \quad \int_\alpha^\beta g(t)\, h(t)\, dt = \sum_{n=0}^{N-1} s_n + R_N,$$

where

$$(6) \quad s_n = (-1)^n \left[g_n(\beta)\, h_{-n-1}(\beta) - g_n(\alpha)\, h_{-n-1}(\alpha)\right]$$

$$(7) \quad R_n = (-1)^n \int_\alpha^\beta g_n(t)\, h_{-n}(t)\, dt,$$

is obtained by repeated integrations by parts. If (α, β) is a finite interval, (5) is valid provided that g is N times continuously differentiable and h is integrable; if (α, β) is an infinite interval then all the integrals involved, and also the limits of $g_n(t)\, h_{-n-1}(t)$ as $t \to \alpha, \beta$, must be assumed to exist.

If g is $N + 1$ times continuously differentiable, a further integration by parts shows that

$$(8) \quad R_N = s_N + R_{N+1}$$

and in certain cases it is possible to use this relation to compare the "remainder", R_N, with the first "neglected term" s_N.

If g and h are real, and $g_N\, h_{-N}$ and $g_{N+1}\, h_{-N-1}$ have constant and equal signs for $a \le t \le \beta$, then R_N has the same sign as s_N, and $|R_N| \le |s_N|$. The proof follows from (8) on noting that in this case R_N and R_{N+1} have opposite signs, and hence R_N and s_N must have the same sign, and

$$|R_N| = \Big|\, |s_N| - |R_{N+1}| \,\Big| .$$

If g is real, $|h_{-N-1}|$ is an increasing function of t, and g_N, g_{N+1} have constant and equal signs for $a \le t \le \beta$, or else if g is real, $|h_{-N-1}|$ is a decreasing function of t, and g_N, g_{N+1} have constant and opposite signs for $a \le t \le \beta$, then $|R_N| \le 2|s_N|$.

We shall prove this result when $|h_{-N-1}|$ is an increasing function of t and $g_N \ge 0$, $g_{N+1} \ge 0$. Then

$$\begin{aligned} |R_{N+1}| &\le |h_{-N-1}(\beta)|\,(g_N(\beta) - g_N(a)) \\ &\le |h_{-N-1}(\beta)\, g_N(\beta)| - |h_{-N-1}(a)\, g_N(a)| \\ &\le |h_{-N-1}(\beta)\, g_N(\beta) - h_{-N-1}(a)\, g_N(a)| \end{aligned}$$

and hence $|R_{N+1}| \le |s_N|$. From (8) we then have the desired result. If $g_N \le 0$ and $g_{N+1} \le 0$, replace g by $-g$. The result for decreasing $|h_{-N-1}|$ follows on replacing x by $-x$.

As an application of these results, let us consider $f(x)$ as defined by (1). If $x > 0$, put $g(t) = (1 + xt)^{-1}$, $h(t) = e^{-t}$, $h_{-m}(t) = (-1)^m\, e^{-t}$. In this case $g_m\, h_{-m} \ge 0$ for all $t \ge 0$, and hence $0 \le (-1)^m\, R_m \le (-1)^m\, s_m$. If x is complex our results do not apply. However, if in (1) we replace t by t/x and accordingly set $g = (1 + t)^{-1}$, $h = x^{-1}\exp(-t/x)$, and then let x become complex, with $\operatorname{Re} x > 0$, then g_m and g_{m+1} are of constant and opposite sign, and $|h_{-m-1}|$ is a decreasing function of t, for $t \ge 0$, and hence $|R_m| \le 2|s_m|$. (Actually, in this case it is easy to prove from (7) that $|R_m| \le |s_m|$.)

Let us suppose now that the integrand in (5), and hence also s_n, R_N, depend on a variable x. If $\{s_n\}$ is an asymptotic sequence, and if in addition we are able to prove $R_N = O(s_N)$ by one of the above results, or in some other manner, then (5) provides an asymptotic expansion of the integral to N terms. For instance, in the case of (2), $\{x^n\}$ is an asymptotic sequence for $x \to 0$; we have proved $|R_N| \le 2|s_N|$ for any N and $\operatorname{Re} x > 0$, and hence (2) is an asymptotic expansion of $f(x)$ as defined in S_Δ, $\Delta > 0$. [Actually, we proved in the Introduction, and could prove from the last integral in (2), that the asymptotic expansion holds in the more extended region S_Δ, $\Delta > -\pi/2$.]

An asymptotic sequence $\{s_n\}$ often occurs if $h(t) = k(xt)$. Denoting by $k_{-m}(u)$ the m-th repeated integral of $k(u)$ with respect to u, we obtain from (5),

$$(9)\quad \int_\alpha^\beta g(t)\, k(xt)\, dt = \sum_{n=0}^{N-1} (-1)^n x^{-n-1} [g_n(\beta)\, k_{-n-1}(\beta x) - g_n(\alpha)\, k_{-n-1}(\alpha x)] + R_N.$$

If the $k_{-n}(u)$ are bounded, and the $g_n(\beta)\, k_{-n-1}(\beta x) - g_n(\alpha)\, k_{-n-1}(\alpha x)$ are bounded away from zero, and if R_N can be estimated as above, then (9) is an asymptotic expansion as $x \to \infty$, the region of x being determined by the estimate of R_N. The apparently more general case $h(t) = k[x\phi(t)]$ can be reduced to the former case by breaking up (α, β) into sub-intervals in which $\phi(t)$ is monotonic, and introducing $\phi(t)$ as a new variable in each of these sub-intervals. In applying one or the other of the above criteria for the estimate of R_N, we then need information about derivatives of the form

$$\frac{d^m}{d\phi^m}\left[\frac{g(t)}{\phi'(t)}\right].$$

If g and ϕ have derivatives of constant or alternating signs, this information can be obtained without explicit computation from results on absolutely and completely monotonic f mctions, see, for instance, Widder (1941) Chapter IV. General theorems of this nature have been obtained by van der Corput and Franklin (1951). The most important application of these methods is to integrals of the form

$$\int_a^b g(t)\, e^{xh(t)}\, dt, \qquad \int_a^b g(t) \begin{smallmatrix}\cos\\ \sin\end{smallmatrix} [xh(t)]\, dt.$$

More general results involving functions of the form $g(t) = (t-a)^{-\lambda} g_1(t)$ with $g_1(t)$ possessing continuous derivatives were obtained by van der Corput (1934).

2.2. Laplace integrals

Integrals of the form

$$(1)\quad f(x) = \int_0^\infty e^{-xt}\, \phi(t)\, dt \equiv \mathfrak{L}\{\phi\}$$

are called *Laplace integrals*. Such integrals occur in the solution of differential equations by definite integrals, and in many other problems. The infinite integral in (1) will be interpreted as the limit of $\int_0^T$ as $T \to \infty$, and it will always be assumed that $\phi(t)$ is integrable over any interval $0 \le t \le T$, $T < \infty$. A function ϕ will be said to belong to $L(x_0)$ if the integral in (1) exists, in the sense mentioned, for $x = x_0$. It is known (Widder, 1941, Chapter II) that for a function ϕ in $L(x_0)$, $\mathfrak{L}\{\phi\}$ exists, and represents an analytic function of x, in the half-plane $\operatorname{Re}(x - x_0) > 0$. In particular, if $\phi(t)$ is integrable over any interval $0 \le t \le T$, $T < \infty$, and $\phi(t) = O(e^{at})$ for some constant a, as $t \to \infty$, then $\mathfrak{L}\{\phi\}$ exists as an (absolutely convergent) infinite integral, and represents an analytic function of x, in the half-plane $\operatorname{Re}(x - a) > 0$.

Under certain circumstances, the asymptotic behavior of $f(x)$ as $x \to \infty$ may be investigated by integrations by parts. *If $\phi(t)$ is N times continuously differentiable for $0 \le t \le a$ and belongs to $L(x_0)$ for some x_0 then*

$$(2) \quad f(x) \sim \Sigma\, \phi^{(n)}(0)\, x^{-n-1}$$

to N terms, uniformly in arg x, *as $x \to \infty$ in S_Δ*, $\Delta > 0$. To prove this, let $\operatorname{Re}(x - x_0) > 0$, and let

$$(3) \quad f(x) = \int_0^a e^{-xt}\phi(t)\,dt + \int_a^\infty e^{-xt}\phi(t)\,dt.$$

The second integral exists and may be integrated by parts to give

$$I_2 = \frac{1}{x - x_0}\int_a^\infty e^{-(x-x_0)t}\,\psi(t)\,dt,$$

where

$$\psi(t) = \int_a^t e^{-x_0 u}\,\phi(u)\,du$$

is a bounded function, say $|\psi| \le A$ for $t \ge a$, so that

$$|I_2| \le \frac{A\,e^{-(\rho-\rho_0)a}}{|x - x_0|\;|\rho - \rho_0|} = O(e^{-\rho a})$$

as $x = \rho + i\sigma \to \infty$ in S_Δ, $\Delta > 0$. In the first integral use 2.1(5) with $g = \phi(t)$, $h_{-n} = (-x)^{-n} e^{-xt}$.

$$s_n = [\phi^{(n)}(0) - \phi^{(n)}(a)\, e^{-xa}]\, x^{-n-1}$$
$$= \phi^{(n)}(0)\, x^{-n-1} + O(e^{-\rho a})$$

and

$$R_N = x^{-N} \int_0^a \phi^{(N)}(t)\, e^{-xt}\, dt \,.$$

Here $\phi^{(N)}(t)$, being continuous, is bounded, say $|\phi^{(N)}(t)| \le B$ for $0 \le t \le a$, and

$$|R_N| \le B\,|x|^{-N} \rho^{-1} \le B\,|x|^{-N-1} \operatorname{cosec} \Delta = O(x^{-N-1}),$$

uniformly in arg x, as $x \to \infty$ in S_Δ, $\Delta > 0$. The proof is completed by noting that $O(e^{-\rho a}) = o(x^{-N})$, uniformly in arg x, as $x \to \infty$ in S_Δ, $\Delta > 0$.

A considerable extension of the last theorem may be based on the following LEMMA. Let $\phi(t)$ and $\psi(t)$ be in $L(x_0)$ for some x_0, $\psi(t) > 0$, $f = \mathfrak{L}\{\phi\}$, $g = \mathfrak{L}\{\psi\}$. *If* $e^{a\rho}\, g(\rho) \to \infty$ *as* $\rho \to \infty$ *for each* $a > 0$, *and if* $\phi(t) = o(\psi(t))$ *as* $t \to 0$, *then* $f(x) = o(g(\rho))$, *uniformly in* arg x, *as* $x = \rho + i\sigma \to \infty$ *in* S_Δ, $\Delta > 0$.

Proof: Given $\epsilon > 0$, there exists an $a > 0$ so that $|\phi| \le \epsilon\psi$ for $0 < t \le a$. With this a, decompose $\mathfrak{L}\{\phi\}$ as in (3). As in the proof of the previous theorem, the second integral is $O(e^{-a\rho})$ as $\rho \to \infty$, and

$$\left|\int_0^a e^{-xt}\, \phi(t)\, dt\right| \le \epsilon \int_0^a e^{-\rho t}\, \psi(t)\, dt = \epsilon\, g(\rho).$$

Thus

$$\frac{|f(x)|}{g(\rho)} \le \epsilon + O\left(\frac{e^{-a\rho}}{g(\rho)}\right),$$

and this is $\le 2\epsilon$ for sufficiently large ρ. The uniformity in arg x follows from the remark that $|x|/\rho \le \operatorname{cosec} \Delta$ in S_Δ.

From this lemma the following theorem can be deduced. For $n = 1, \ldots, N$, let $\psi_n(t)$ be in $L(x_0)$ for some x_0, $\psi_n(t) > 0$ for $t > 0$, and $g_n = \mathfrak{L}\{\psi\}$. *If* $\{\psi_n\}$ *is an asymptotic sequence for* $t \to 0$, *and* $e^{a\rho}\, g_n(\rho) \to \infty$ *as* $\rho \to \infty$ *for each* $a > 0$ *and each* n, *then* $\{g_n(\rho)\}$ *is an asymptotic sequence for* $\rho \to +\infty$; *and if under these circumstances* $\phi(t)$ *is in* $L(x_0)$ *and*

$$\phi(t) \sim \sum a_n \psi_n(t) \quad \text{to } N \text{ terms as } t \to 0,$$

then

$$f(\rho) \sim \sum a_n g_n(\rho) \quad \text{to } N \text{ terms as } \rho \to +\infty.$$

If in addition for each $n = 1, \ldots, N$, $g_n(\rho)/g_n(x)$ *is bounded in* S_Δ *for sufficiently large* $|x|$, *then also* $\{g_n(x)\}$ *is an asymptotic sequence, and* $f(x) \sim \sum a_n g_n(x)$ *to N terms, uniformly in* arg x, *as* $x \to \infty$ *in* S_Δ, $\Delta > 0$.

Proof: From the lemma we have $g_{n+1}(\rho) = o(g_n(\rho))$, and hence $\{g_n(\rho)\}$ is an asymptotic sequence for $\rho \to \infty$. To prove the asymptotic expansion for $\rho \to \infty$, replace ϕ by

$$\phi - \sum_{n=1}^{N} a_n \psi_n,$$

and ψ by ψ_N, in the lemma. When $x \to \infty$ in S_Δ, we have from the lemma and the additional assumption on $g_n(x)$ that

$$g_{n+1}(x) = o(g_n(\rho)) = \frac{g_n(\rho)}{g_n(x)}\, o(g_n(x)) = o(g_n(x)),$$

and hence $\{g_n(x)\}$ is an asymptotic sequence for $x \to \infty$ in S_Δ. The proof of the asymptotic expansion is the same as in the previous case.

The most important particular case of our general theorem is

$$\psi_n = t^{\lambda_n - 1} \qquad 0 < \lambda_1 < \cdots < \lambda_N.$$

All conditions of the theorem are satisfied: in particular, for

$$g_n(x) = \Gamma(\lambda_n)\, x^{-\lambda_n}$$

we have

$$\frac{g_n(\rho)}{|g_n(x)|} \le \left(\frac{|x|}{\rho}\right)^{\lambda_n} \le (\operatorname{cosec} \Delta)^{\lambda_n}$$

for x in S_Δ, $\Delta > 0$. We then obtain the following theorem on asymptotic series of powers.

Let $0 < \lambda_1 < \lambda_2 < \cdots$. *If* $\phi(t)$ *is in* $L(x_0)$ *for some* x_0 *and*

$$\phi \sim \Sigma\ a_n t^{\lambda_n - 1} \quad \text{to } N \text{ terms as } t \to 0,$$

then

$$f \sim \Sigma\, \Gamma(\lambda_n)\, a_n\, x^{-\lambda_n} \quad \text{to } N \text{ terms, uniformly in arg } x,$$

as $x \to \infty$ in S_Δ, $\Delta > 0$.

Other notable examples of asymptotic sequences to which the general theorem applies are

$$(4) \quad \psi_n = (1 - e^{-t})^{n-1}, \qquad g_n = \frac{(n-1)!}{x(x+1)\cdots(x+n-1)}$$

$$(5) \quad \psi_n = (e^{t} - 1)^{n-1}, \qquad g_n = \frac{(n-1)!}{x(x-1)\cdots(x-n+1)}$$

$$(6) \quad \psi_n = \left(2 \sinh \frac{t}{2}\right)^{2n-2}, \qquad g_n = \frac{(2n-2)!}{(x-n+1)(x-n+2)\cdots(x+n-1)}.$$

In the case of asymptotic power series or in the case of (4), N may be finite or infinite, in the cases of (5) and (6) N must be finite.

The result obtained by integrations by parts is a particular case of of the above theorem for asymptotic series of powers. If $\phi(t)$ is N times continuously differentiable for $0 \le t \le a$, then it can be proved by the mean value theorem of differential calculus that $\phi(t) \sim \Sigma\ \phi^{(n)}(0)\ t^n/n!$ to N terms as $t \to 0$, and then (2) follows from the theorem on asymptotic power series.

In many cases it is possible to extend the region of validity of the asymptotic expansion to an S_Δ with $\Delta < 0$. If $\phi(t)$ is an analytic function of t which is regular in S_θ, and is $O(e^{at})$ for some constant a, as $t \to \infty$ in S_θ, then it is permissible to rotate the path of integration in (1) to any ray in S_θ, and by this means $f(x)$ may be continued analytically to some region which contains the sector $-\pi + \theta < \arg(x - a) < \pi - \theta$ [see, for instance, Doetsch (1950, p. 362 ff.)]. If $\{\psi_n\}$ has suitable properties along each ray, and $\phi \sim \Sigma\, a_n\, \psi_n$ as $t \to 0$ in S_θ, then $f \sim \Sigma\, a_n\, g_n$ as $x \to \infty$ in S_Δ where $\Delta > \theta - \pi/2$. We shall formulate a precise theorem for asymptotic series of powers.

If $\phi(t)$ is a regular function of t in S_θ, $\phi = O(e^{at})$, uniformly in arg t, *for some a, as $t \to \infty$ in S_θ,*

$$\phi \sim \Sigma\, a_n t^{\lambda_n - 1} \quad \text{to } N \text{ terms, uniformly in arg } t, \text{ as } t \to 0 \text{ in } S_\theta,$$

where

$$0 < \lambda_1 < \lambda_2 < \cdots < \lambda_N,$$

then $f(x)$ exists at least in the sector

$$-\pi + \theta < \arg(x - a) < \pi - \theta,$$

and

$$f \sim \Sigma\, a_n \Gamma(\lambda_n)\, x^{-\lambda_n} \quad \text{to } N \text{ terms, uniformly in arg } x,$$

as $x \to \infty$ in S_Δ, $\Delta > \theta - \pi/2$.

The particular case of this result in which the λ_n form an arithmetic progression, and ϕ is represented by a convergent infinite series

$$\Sigma\, a_n t^{\lambda_n - 1}$$

for sufficiently small $|t|$ in S_θ is known as *Watson's lemma*; it is sufficient for many applications.

Throughout this section we investigated the behavior of (1) for large x. Similar methods may be used for the investigation of $f(x)$ as $x \to x_0$. For the basic lemma see Erdélyi (1947), and for some of the most important results see Doetsch (1950) Chapter 13 and Widder (1941) Chapter V.

2.3. Critical points

We have seen in the last two sections that under certain circumstances the asymptotic behavior of integrals is determined by the behavior of the integrands at certain distinguished points, the end-points of the interval in the cases considered in the preceding sections. Such distinguished points have been called critical points by van der Corput (1948). There is no general theory of critical points but a few types of such points, and the methods adapted to deal with them may be described as follows.

First let us consider an integral of the form

$$(1)\quad \int_{\alpha}^{\beta} g(t)\, e^{xh(t)}\, dt\,,$$

where x is a large positive parameter and $h(t)$ is real. If $h(t)$ has a maximum at $t = \tau$, and $h(t) < h(\tau)$ when $t \neq \tau$, then for large x, the modulus of the integrand will have a sharp maximum at a point very near τ, and most of the contribution to the integral will arise from the immediate vicinity of this maximum. The integral can be evaluated approximately by expanding both g and h in the neighborhood of $t = \tau$. This is the central idea of *Laplace's method* (sec. 2.4). We have encountered such a case in sec. 2.2 where $h(t) = -t$, $0 \leq t < \infty$, and $h(t)$ has a maximum at $t = 0$. Accordingly, we evaluated Laplace integrals asymptotically by expanding $g(t)$ for small values of t.

If x and $h(t)$ are complex, and $g(t)$ and $h(t)$ are analytic functions of t, then it is often possible to deform the path of integration so that it passes through one or several points at which $h'(t) = 0$. If τ is such a point, it is a critical point; it is possible to determine that part of the path of integration which passes through τ in such a manner that $x[h(t) - h(\tau)]$ is real along the path, and the integral can then be evaluated by an adaptation of Laplace's method. This is Riemann's *method of steepest descents* (sec. 2.5).

Next let us turn to integrals of the form

$$(2)\quad \int_{\alpha}^{\beta} g(t)\, e^{ixk(t)}\, dt$$

where we again assume that x is a large positive parameter, and $k(t)$ is a real function. In general, the rapid oscillations of $\exp[ixk(t)]$ will tend to cancel large contributions to the integral, but this cancellation will not occur at the end-points, or at the stationary points of $k(t)$. If $k(t)$ has no stationary points in the interval $\alpha < t < \beta$, integration by parts (sec. 2.1) will in general give a good approximation. Stokes' *method of stationary phase* (sec. 2.9) appraises the contribution of a stationary point, τ, to the integral by expanding g and h in the neighborhood of this point.

The method of stationary phase has been extended by van der Corput (1936) to integrals of the form (1) where $xh(t)$ may be complex (instead of being imaginary, as in the method of stationary phase). According to van der Corput, the critical points in this case are those points, τ, at which $x^{1/2}\, h'(t)\, [h''(t)]^{-1/2}$ is real, while the imaginary part of this function changes its sign when t passes through τ. (In the case of (2), $h = ik$, k is real, and the only points at which

$$x^{1/2}\, h'\, (h'')^{-1/2} = (ix)^{1/2}\, k'\, (k''.)^{-1/2}$$

is real are the stationary points of k.)

2.4. Laplace's method

In the integral

$$(1) \quad f(x) = \int_a^\beta g(t)\, e^{xh(t)}\, dt$$

let $h(t)$ be a real function of the real variable t, while $g(t)$ may be real or complex, and let x be a large positive variable. According to Laplace, the major contribution to the value of the integral arises from the immediate vicinity of those points of the interval $a \le t \le \beta$ at which $h(t)$ assumes its largest value. If $h(t)$ has a finite number of maxima, we may break up the integral in a finite number of integrals so that in each integral $h(t)$ reaches its maximum at one of the end-points and at no other point. Accordingly, we shall assume that $h(t)$ in (1) reaches its maximum at $t = a$, and that $h(t) < h(a)$ for $a < t \le \beta$.

Assuming g continuous and h twice continuously differentiable, $h'(a) = 0$, $h''(a) < 0$, Laplace introduced a new variable u by the substitution $h(a) - h(t) = u^2$. $h'(t)$ will be negative in $a < t < a + \eta$ for some sufficiently small η. As $x \to \infty$,

$$f(x) \sim \int_a^{a+\eta} g(t)\, e^{xh(t)}\, dt = -\int_0^U \quad 2u\, \frac{g(t)}{h'(t)}\, \{\exp x[h(a) - u^2]\}\, du$$

where $U = [h(a) - h(a + \eta)]^{1/2} > 0$. Since only the neighborhood of $u = 0$ matters, we may replace $g(t)$ approximately by $g(a)$, and $u/h'(t)$ by $-[-2h''(a)]^{-1/2}$, which is the limit of $u/h'(t)$ as $t \to a$, and obtain

$$f(x) \sim \left[\frac{-2}{h''(a)}\right]^{1/2} g(a) \int_0^U \{\exp[-xu^2 + xh(a)]\}\, du$$

By the same argument, we may extend the integration to $u = \infty$ and finally obtain Laplace's result

(2) $$f(x) \sim g(a)\, e^{xh(a)} \left[\frac{-\pi}{2xh''(a)}\right]^{\frac{1}{2}} \qquad x \to \infty.$$

Later, Burkhardt (1914) and Perron (1917) showed that the same result can be proved by expanding g and h in the neighborhood of a. Copson (1946) reproduces a simple proof of Pólya and Szegö, and Widder (1941, Chapter VII) gives a more sophisticated proof under more general conditions. Further extensions of Laplace's formula were obtained by Hsu (1949 a, b; 1951 a, b), Levi (1946) and Rooney (1953). Laplace's method has been applied to integrals depending on two large variables by Fulks (1951) and Thomsen (1954), and to double and multiple integrals by Hsu (1948 a, b; 1951 c) and Rooney (1953).

The following extension of Laplace's result will be derived from our discussion of Laplace integrals. *Let g and h be functions on the interval (a, β) for which the integral* (1) *exists for each sufficiently large positive x, let h be real, continuous at $t = a$, continuously differentiable for $a < t \le a + \eta$, $\eta > 0$, and such that $h' < 0$ for $a < t \le a + \eta$, $h(t) \le h(a) - \epsilon$, $\epsilon > 0$, for $a + \eta \le t \le \beta$; suppose that $h'(t) \sim -a(t-a)^{\nu-1}$ and $g(t) \sim b(t-a)^{\lambda-1}$* as $t \to a$, $\lambda > 0$, $\nu > 0$: *then*

(3) $$f(x) = \int_a^\beta g(t)\, e^{xh(t)}\, dt \sim \frac{b}{\nu} \quad \Gamma\left(\frac{\lambda}{\nu}\right)\left(\frac{\nu}{ax}\right)^{\lambda/\nu} e^{xh(a)} \qquad x \to \infty.$$

We first note that

(4) $$\left|\int_{a+\eta}^{\beta} g(t)\, e^{xh(t)}\, dt\right| \le \exp\{x[h(a) - \epsilon]\} \int_{a+\eta}^{\beta} |g(t)|\, dt$$
$$= o\left[x^{-\lambda/\nu}\, e^{xh(a)}\right] \qquad x \to \infty.$$

In the interval $(a,\ a + \eta)$ we introduce a new variable $u = h(a) - h(t)$, set $U = h(a) - h(a+\eta) > 0$, $k(u) = -g(t)/h'(t)$ and obtain

(5) $$\int_a^{a+\eta} g(t)\, e^{xh(t)}\, dt = e^{xh(a)} \int_0^U k(u)\, e^{-xu}\, du.$$

Now,

$$u = h(a) - h(t) = -\int_a^t h'(\tau)\,d\tau \sim \frac{a}{\nu}(t-a)^{\nu} \qquad \text{as } t \to a$$

and hence

$$t - a \sim \left(\frac{u\nu}{a}\right)^{1/\nu} \qquad \text{as } u \to 0.$$

Also,

$$-\frac{g(t)}{h'(t)} \sim \frac{b}{a}(t-a)^{\lambda-\nu} \qquad \text{as } t \to a$$

so that

$$(6) \quad k(u) \sim \frac{b}{a}\left(\frac{u\nu}{a}\right)^{\lambda/\nu - 1} \qquad \text{as } u \to 0.$$

By the results of the preceding section on the asymptotic behavior of Laplace integrals it follows from (5) and (6) that

$$(7) \quad \int_a^{a+\eta} g(t)\,e^{xh(t)}\,dt \sim \frac{b}{\nu}\left(\frac{\nu}{a}\right)^{\lambda/\nu}\Gamma\left(\frac{\lambda}{\nu}\right)x^{-\lambda/\nu}\,e^{xh(a)}$$

$$\text{as } x \to \infty,$$

and (4) and (7) prove (3). Moreover, both (4) and (7), and hence also (3), remain true if x is a complex variable and $x \to \infty$ in S_Δ, $\Delta > 0$.

A further extension of (3) leads to an asymptotic expansion of $f(x)$. In the following formulas $n = 0, 1, \dots, N-1$. If

$$(8) \quad -h'(t) \sim \Sigma\, a_n (t-a)^{\nu+n-1}, \quad g(t) \sim \Sigma\, b_n (t-a)^{\lambda+n-1}$$

to N terms as $t \to a$, then there is an expansion

$$(9) \quad -\frac{g(t)}{h'(t)} \sim \Sigma\, c_n (t-a)^{\lambda-\nu+n} \qquad \text{to } N \text{ terms as } t \to a$$

and the c_n may be computed by formal division. Also

$$(10) \quad u = -\int_a^t h'(\tau)\,d\tau \sim \sum \frac{a_n}{\nu+n}(t-a)^{\nu+n} \qquad \text{to } N \text{ terms as } t \to a.$$

From this last expansion it may be shown that $t-a$ possesses an asymptotic power series expansion in powers of $u^{1/\nu}$, this asymptotic power series may be substituted in (9), and leads to an asymptotic expansion of the form

$$(11)\quad k(u) \sim \Sigma \gamma_n u^{(\lambda+n-\nu)/\nu} \quad \text{to } N \text{ terms as } u \to \infty.$$

With (11) instead of (6), an asymptotic expansion to N terms is obtained for (5); (4) may be strengthened to $o\,(x^{-(\lambda+N)/\nu}\, e^{xh(a)})$; and we have

$$(12)\quad f(x) \sim e^{xh(a)}\, \Sigma \gamma_n \Gamma\left(\frac{\lambda+n}{\nu}\right) x^{-(\lambda+n)/\nu}$$

to N terms as $x \to \infty$ in S_Δ, $\Delta > 0$. The coefficients γ_n may be computed by formal substitution according to the scheme described above.

There is an alternative procedure for the computation of the γ_n which avoids the necessity for inverting the asymptotic series (10) to obtain the expansion of $t-a$ in powers of $u^{1/\nu}$. From (10),

$$h(t) = h(a) - \frac{a_0}{\nu}(t-a)^\nu + h_1(t)$$

where

$$h_1(t) = -\sum_{n=1}^{N-1} \frac{a_n}{\nu+n}(t-a)^{\nu+n} + o\,((t-a)^{\nu+N-1}).$$

We now write

$$\int_a^{a+\eta} g(t)\, e^{xh(t)}\, dt = e^{xh(a)} \int_a^{a+\eta} l(t) \exp\left[-\frac{a_0}{\nu}\, x(t-a)^\nu\right] dt,$$

expand

$$l(t) = g(t) \exp[xh_1(t)]$$

formally in powers of $(t-a)$, and integrate term-by-term to obtain (12). It is also possible to construct a proof of (12) along these lines.

2.5. The method of steepest descents

We again consider the integral

$$(1)\quad f(x) = \int_a^\beta g(t)\, e^{xh(t)}\, dt,$$

in which we now assume x to be a large complex variable, g and h to be analytic functions of the complex variable t, and the integral to be taken along some path in the complex t plane. This integral may be evaluated asymptotically by the method of steepest descents, which was originated by Riemann and developed by Debye. Copson (1946) gives a detailed description of this method together with references and several examples.

Those points of the t plane at which $h'(t) = 0$ will be called *saddle points or cols*. The surface representing $|\exp[xh(t)]|$ as a function of Re t and Im t will be called the *relief* of e^{xh}: on this surface cols will be "saddles", and the most convenient trail from one "valley" to the other will lead over one or several saddles. Let τ be a col: if $h'(\tau) = h''(\tau) = \cdots = h^{(m)}(\tau) = 0$ and $h^{(m+1)}(\tau) \neq 0$, we call τ a *col* (or saddle point) *of order m*. In the t-plane, curves along which Re $xh(t)$ is constant are called *level curves*: along such curves e^{xh} has a constant modulus (they are contour lines of the relief), and the phase of e^{xh} changes as rapidly as possible. Those curves along which Im $xh(t)$ is constant are called *steepest paths*: along such curves e^{xh} has a constant phase, and the modulus of e^{xh} changes as rapidly as possible (they are gradient lines of the relief). At a col, τ, of order m, $m+1$ level curves intersect at equal angles, and their angles are bisected by $m+1$ steepest paths: along each of the latter curves $|e^{xh(t)}|$ has a stationary point at τ.

The method of steepest descents consists in deforming the path of integration so as to make it coincide as far as possible with arcs of steepest paths. If α and β lie on steepest arcs through cols, for instance if α and β are singularities of $h(t)$, then the path of integration may be deformed so as to consist entirely of steepest paths through cols; otherwise two steepest arcs may occur which do not pass through cols. This latter case may be described by reference to the relief by saying that we first descend along a gradient line to a singularity and them climb the saddle along another gradient line. In any event, Re $xh(t)$ is monotonic along any steepest path (except at saddles), and Laplace's method may be used to evaluate the integral asymptotically. The asymptotic expansions of g and h needed for the application of the theorem in the preceding section are the Taylor expansions of g and h around that point of the steepest path at which Re $xh(t)$ is a maximum (this is often the col). The inversion of the series 2.4(10) may be effected by Lagrange's expansion (see, for instance, Copson (1935), p. 123-125).

Meijer (1933 a,b) has shown that numerical bounds for the error term may be obtained by using Lagrange's expansion with a remainder, and he has also shown that in some cases recurrence relations for the coefficients may also be obtained.

We shall consider several examples of the application of the method of steepest descents: these are taken from Copson (1946).

2.6. Airy's integral

We shall investigate the asymptotic behavior of

$$(1)\quad Ai(z) = \frac{1}{\pi}\int_0^\infty \cos\left(\frac{1}{3}s^3 + zs\right) ds$$

for large positive values of z. With

$$(2)\quad s = z^{1/2}t, \qquad x = z^{3/2}$$

we obtain

$$(3)\quad Ai(x^{2/3}) = \frac{x^{1/3}}{2\pi}\int_{-\infty}^{\infty} \exp\left[ix\left(\frac{1}{3}t^3 + t\right)\right] dt,$$

and the method of steepest descents can be applied to the integral in (3). In (3), t may be envisaged as a complex variable of integration. The path of integration (for $x > 0$) is the real t axis, but it can be deformed into any curve which begins at infinity in the sector $2\pi/3 < \arg t < \pi$ and ends at infinity in the sector $0 < \arg t < \pi/3$. Here

$$h(t) = i\left(\frac{1}{3}t^3 + t\right), \qquad h'(t) = i(t^2 + 1),$$

and the cols are the zeros of $h'(t)$, i.e., the points $t = \pm i$. The steepest paths are determined by $\operatorname{Im} h(t) = \text{const}$. We set $t = \xi + i\eta$ and obtain

$$\operatorname{Im} h(t) = \frac{1}{3}\xi^3 - \xi\eta^2 + \xi, \qquad \operatorname{Im} h(\pm i) = 0,$$

so that the equation of the steepest paths is

$$(4)\quad \xi(\xi^2 - 3\eta^2 + 3) = 0.$$

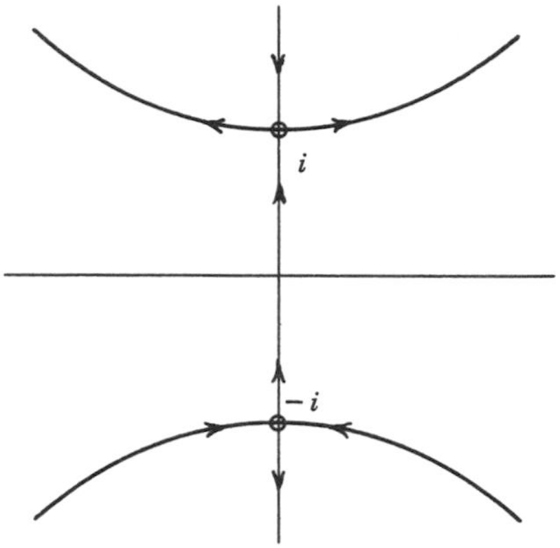

This equation represents a degenerate cubic consisting of the imaginary axis and of the two branches of a hyperbola. In the figure, arrows indicate the direction in which Re $h(t)$ decreases. The asymptotes of the hyperbola are the lines $\xi \pm \eta\sqrt{3} = 0$, and clearly, the path of integration in (3) can be deformed into the upper branch of the hyperbola, and runs from $\infty \cdot \exp(5\pi i/6)$ to $\infty \cdot \exp(i\pi/6)$. With this path, the integral in (3) can be seen to be convergent whenever Re $x > 0$.

We now write

$$(5)\quad 2\pi x^{-1/3} Ai(x^{2/3}) = \int_i^{\infty \cdot \exp(i\pi/6)} - \int_i^{\infty \cdot \exp(5i\pi/6)} e^{xh(t)}\,dt$$

$$= I_1 - I_2$$

and evaluate $I_{1,2}$ by Laplace's method. In both integrals, $h(t) - h(i)$ is real and reaches its maximum at $t = i$; also $h(t) - h(i)$ is a decreasing function. We introduce a new variable u by

$$(6)\quad u = h(i) - h(t) = -\frac{2}{3} - i\left(\frac{1}{3}t^3 + t\right) = (t-i)^2 - \frac{1}{3}i(t-i)^3.$$

From (6)

$$(7) \quad \pm u^{1/2} = (t - i)\left[1 - \frac{1}{3} i(t - i)\right]^{1/2},$$

where $u^{1/2}$ is the positive square root, $[\cdots]^{1/2}$ is that value which reduces to 1 at $t = i$, and the upper sign in (7) holds for I_1, the lower sign for I_2. It follows from Lagrange's theorem that sufficiently near to the col, $t-i$ possesses an expansion of the form $t - i = \sum b_n (\pm u^{1/2})^n$ where nb_n is the coefficient of $(t - i)^{n-1}$ in the expansion of $[1 - i(t - i)/3]^{-n/2}$ in powers of $t - i$. In this manner the expansions

$$(8) \quad t - i = \sum_{n=1}^{\infty} \frac{i^{n-1}\Gamma(3n/2 - 1)}{n!\Gamma(n/2)\, 3^{n-1}} (\pm u^{1/2})^n$$

are obtained where the upper or lower sign holds in I_1 and I_2 respectively. Now

$$e^{-xh(i)} I_{1,2} = \int_0^\infty e^{-xu} \frac{dt}{du}\, du,$$

and according to sec. 2.4 the asymptotic expansions of $I_{1,2}$ are obtained by substituting (8) in dt/du and then integrating term-by-term. Thus

$$e^{2x/3} I_{1,2} = \int_0^\infty e^{-xu} \sum_{n=1}^{\infty} \frac{(\pm 1)^n i^{n-1}\Gamma(3n/2 - 1)}{2(n-1)!\Gamma(n/2)\, 3^{n-1}} u^{n/2-1}\, du$$

$$\sim \sum_{n=1}^{\infty} \frac{(\pm 1)^n i^{n-1}\Gamma(3n/2 - 1)}{2(n-1)!\, 3^{n-1} x^{n/2}}.$$

Substituting this in (5) and expressing the result in terms of z, we obtain after some simplification

$$(9) \quad Ai(z) \sim \frac{1}{2\pi z^{1/4}} \exp\left(-\frac{2}{3} z^{3/2}\right) \sum_{m=0}^{\infty} \frac{\Gamma(3m + \frac{1}{2})}{(2m)!} (-9 z^{3/2})^{-m},$$

and this asymptotic representation holds, uniformly in $\arg z$, as $z \to \infty$ in $|\arg z| \leq \pi/3 - \Delta$, $\Delta > 0$.

2.7. Further examples

We shall now consider two examples where the limits of integration are not singularities, and accordingly, the asymptotic expansions are not obtained by expansions around the col. Also, in the second example, the col is of order two.

First let $x > 0$ and

$$(1) \quad f(x) = \int_0^\infty \exp\left[ix\left(\frac{1}{3}t^3 + t\right)\right] dt.$$

The function $h(t)$ occurring here is the same as in sec. 2.6, the steepest paths are those shown in the figure in sec. 2.6, and it is easily seen that the appropriate path of integration consists of that portion of the imaginary axis from 0 to i and then one-half of the upper branch of the hyperbola. Thus,

$$(2) \quad f(x) = \int_0^i + \int_i^{\infty \cdot \exp(i\pi/6)} e^{xh(t)}\, dt.$$

The asymptotic expansion of the second integral has already been obtained. In the first integral $h(t)$ is real and decreasing as t runs from 0 to i, and we may again use Laplace's method. Accordingly, we set

$$u = h(0) - h(t) = -it\left(1 + \frac{1}{3}t^2\right)$$

and infer from Lagrange's theorem that $-it = \Sigma\, b_n u^n$ where nb_n is the coefficient of $(-it)^{n-1}$ in the expansion of $(1 + t^2/3)^{-n}$ in powers of $-it$. Clearly $b_n = 0$ if n is even and

$$-it = \sum_{m=0}^{\infty} \frac{(3m)!\, u^{2m+1}}{m!\,(2m+1)!\,3^m}.$$

Substituting this in the first integral in (2) and integrating term-by-term,

$$\int_0^i e^{xh(t)}\, dt = i\int_0^{2/3} e^{-xu} \sum_{m=0}^{\infty} \frac{(3m)!\, u^{2m}}{m!\,(2m)!\,3^m}\, du$$

$$\sim i \sum_{m=0}^{\infty} \frac{(3m)!}{m!\,3^m}\, x^{-2m-1}.$$

It is seen from sec. 2.6 that the second integral in (2) is exponentially small in comparison with the first one, and hence the result

$$(3)\quad \int_0^{\infty} \exp\left[ix\left(\frac{1}{3}t^3 + t\right)\right] dt \sim i \sum_{m=0}^{\infty} \frac{(3m)!}{m!\,3^m}\, x^{-2m-1}$$

as $x \to \infty$ in S_Δ, $\Delta > 0$.

Our last example is the integral

$$(4)\quad f(x) = \int_0^1 \exp(ixt^3)\, dt$$

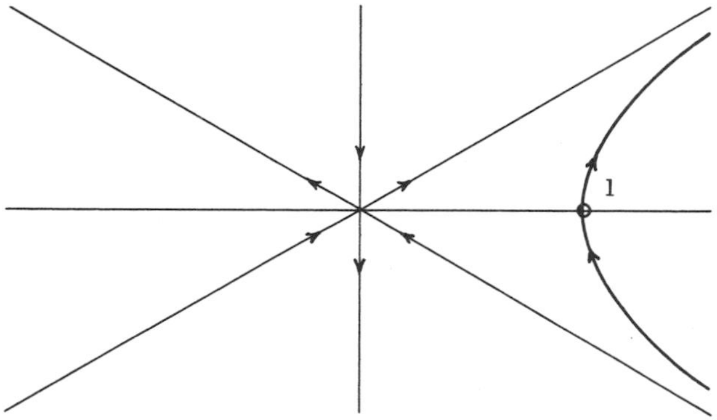

where we take $x > 0$. Here $h(t) = it^3$, and $t = 0$ is a col of order two. The steepest paths through the col are the lines $\operatorname{Im}(it^3) = 0$, that is the lines $\arg t = \pm\pi/6$, $\pm\pi/2$, $\pm 5\pi/6$. In the figure, arrows indicate the direction of decreasing $|\exp(ixt^3)|$ for $x > 0$. None of these steepest paths passes through $t = 1$. With $t = \xi + i\eta$, the equation of the steepest path through $t = 1$ is $\operatorname{Im}(it^3) = 1$, or $\xi^3 - 3\xi\eta^2 = 1$. This is a cubic, and the branch of this cubic passing through $t = 1$ is also indicated in the figure. In order to get from 0 to 1 along steepest paths, we first integrate from 0 to ∞ along the line $\arg t = \pi/6$, and then from ∞ to 1 along the upper half of the branch of our cubic. Accordingly, we set

(5) $f(x) = \int_0^{\infty \cdot \exp(\pi i/6)} - \int_1^{\infty \cdot \exp(\pi i/6)} \exp(ixt^3) dt$

$$= I_1 - I_2$$

In I_1 we set $u = -it^3$ or $t = u^{1/3} e^{\pi i/6}$, where $u^{1/3} > 0$, and obtain

(6) $I_1 = \frac{1}{3} e^{i\pi/6} \int_0^\infty e^{-xu} u^{-2/3} du = \Gamma(4/3) e^{i\pi/6} x^{-1/3}$.

In I_2, we set $u = -i(t^3 - 1)$ or $t = (1 + iu)^{1/3}$ and obtain

$$I_2 = \frac{1}{3} i \int_0^\infty e^{ix - xu} (1 + iu)^{-2/3} du.$$

Expanding $(1 + iu)^{-2/3}$ in the binomial expansion,

(7) $I_2 \sim \frac{1}{\Gamma(-1/3)} e^{ix} \sum_{n=0}^{\infty} \Gamma(n + 2/3)(ix)^{-n-1}$,

and substituting (6) and (7) in (5) we finally have

(8) $\int_0^1 \exp(ixt^3) dt \sim \Gamma(4/3) e^{i\pi/6} x^{-1/3}$

$$- \frac{1}{\Gamma(-1/3)} e^{ix} \sum_{n=0}^{\infty} \Gamma(n + 2/3)(ix)^{-n-1}$$

as $x \to \infty$ in S_Δ, $\Delta > 0$.

The last equation describes the asymptotic behavior of $f(x)$ as $x \to \infty$ in the right half-plane. If $x \to \pm i\infty$, the integrand in (4) is real, and Laplace's method may be applied, and if x is in the left half-plane, we may use the relation

$$f(x) = \overline{f(-\overline{x})}$$

which follows from (4) and in which bars denote complex conjugation.

2.8. Fourier integrals

Integrals of the form

(1) $\int_\alpha^\beta e^{ixt} \phi(t) dt$

are called *Fourier integrals*. We shall always assume that (α, β) is a real interval, and mostly (α, β) will be a finite real interval; and $\phi(t)$ will be an integrable function so that (1) exists for all real x. We shall investigate the asymptotic behavior of (1) as $x \to +\infty$; to obtain the asymptotic behavior as $x \to -\infty$, replace t by $-t$. Unlike in the case of Laplace integrals (sec. 2.2) it seems that repeated integrations by parts is the only effective method for obtaining asymptotic expansions of (1), except in the case of analytic $\phi(t)$ when the method of steepest descents may be used.

First we shall prove: *If $\phi(t)$ is N times continuously differentiable for $\alpha \le t \le \beta$ then*

$$(2)\quad \int_\alpha^\beta e^{ixt}\phi(t)\,dt = B_N(x) - A_N(x) + o(x^{-N}) \qquad \text{as } x \to \infty,$$

where

$$(3)\quad A_N(x) = \sum_{n=0}^{N-1} i^{n-1}\phi^{(n)}(\alpha)\,x^{-n-1}e^{ix\alpha}$$

$$B_N(x) = \sum_{n=0}^{N-1} i^{n-1}\phi^{(n)}(\beta)\,x^{-n-1}e^{ix\beta}$$

and $\phi^{(n)} = d^n\phi/dt^n$. The result remains true when $\alpha = -\infty$ (or $\beta = \infty$) *provided that $\phi^{(n)}(t) \to 0$ as $t \to -\infty$ (or $t \to \infty$) for each $n = 0, 1, \dots, N-1$, and provided further that $\phi^{(N)}(t)$ is integrable over (α, β).* To prove (2), we apply 2.1 (5) with $g = \phi$, $g_n = \phi^{(n)}$, $h = e^{ixt}$, $h_{-n} = (ix)^{-n}e^{ixt}$. For the remainder we obtain

$$R_N = (-ix)^{-N}\int_\alpha^\beta e^{ixt}\phi^{(N)}(t)\,dt$$

and this is $o(x^{-N})$ since the integral approaches zero, as $x \to \infty$, by Riemann's lemma.

We note that $A_N(x) = 0$ if ϕ and its first $N-1$ derivatives vanish at α (for instance, if ϕ vanishes identically in some neighborhood of α), and also that $B_N(x) = 0$ if ϕ and its first $N-1$ derivatives vanish at β (for instance, if ϕ vanishes in some neighborhood of β).

We now turn to Fourier integrals whose integrands have singularities of a simple type at one end point of the interval.

If $\phi(t)$ *is N times continuously differentiable for* $a \le t \le \beta$; $\phi^{(n)}(\beta) = 0$ *for* $n = 0, 1, \dots, N-1$; and $0 < \lambda < 1$; then

(4) $$\int_a^\beta e^{ixt}(t-a)^{\lambda-1}\,\phi(t)\,dt = -A_N(x) + O(x^{-N}) \qquad \text{as } x \to \infty,$$

where

(5) $$A_N(x) = \sum_{n=0}^{N-1} \frac{\Gamma(n+\lambda)}{n!}\, e^{\pi i (n+\lambda-2)/2}\, \phi^{(n)}(a)\, x^{-n-\lambda} e^{ixa}.$$

If $\phi(t)$ *is N times continuously differentiable for* $a \le t \le \beta$; $\phi^{(n)}(a)=0$ *for* $n = 0, 1, \dots, N-1$; *and* $0 < \mu < 1$; *then*

(6) $$\int_a^\beta e^{ixt}(\beta-t)^{\mu-1}\,\phi(t)\,dt = B_N(x) + O(x^{-N}) \qquad \text{as } x \to \infty,$$

where

(7) $$B_N(x) = \sum_{n=0}^{N-1} \frac{\Gamma(n+\mu)}{n!}\, e^{\pi i (n-\mu)/2}\, \phi^{(n)}(\beta)\, x^{-n-\mu} e^{ix\beta}.$$

With $\lambda = 1$, (5) becomes the first equation (3), and with $\mu = 1$, (7) becomes the second equation (3), but the O terms in (4) and (6) give less information than the o term in (2). Instead of $O(x^{-N})$ we could write $o(x^{-N-\lambda+1})$ in (4), and $o(x^{-N-\mu+1})$ in (6), and these latter forms remain valid, and pertinent, when $\lambda = 1$ or $\mu = 1$ respectively.

We shall prove (4): the proof of (6) is similar. In (4) we apply 2.1 (5) with $g(t) = \phi(t)$, $g_n(t) = \phi^{(n)}(t)$, $h(t) = h_0(t) = e^{ixt}(t-a)^{\lambda-1}$, and

(8) $$h_{-n-1}(t) = \frac{(-1)^{n+1}}{n!} \int_t^{i\infty} (u-t)^n (u-a)^{\lambda-1} e^{ixu}\,du$$

$$n = 0, 1, \dots, N-1.$$

In (8) we assume $t > a$, and take a path of integration which lies entirely in the quadrant $0 \le \arg(u-a) \le \pi/2$. The integral converges absolutely, and

$$\frac{d}{dt} h_{-n-1}(t) = h_{-n}(t) \qquad n = 0, 1, \dots, N-1.$$

If we take the ray $u = t + i\sigma$, $\sigma \geq 0$, as the path of integration, we have $|u - a| \geq t - a$, $|u - a|^{\lambda-1} \leq (t - a)^{\lambda-1}$ for $0 < \lambda \leq 1$, and hence

$$|h_{-n-1}(t)| \leq \frac{(t-a)^{\lambda-1}}{n!} \int_t^{t+i\infty} |u - t|^n \, |e^{ixu} \, du|$$

Substituting $u = t + i\sigma$ we have

$$(9) \quad |h_{-n-1}(t)| \leq (t-a)^{\lambda-1} x^{-n-1} \qquad t > a, \ x > 0.$$

Also, from (8),

$$h_{-n-1}(a) = \frac{(-1)^{n+1}}{n!} \int_a^{a+i\infty} (u-a)^{n+\lambda-1} e^{ixu} \, du$$

and, with $u = a + i\sigma$, $\sigma \geq 0$,

$$(10) \quad h_{-n-1}(a) = (-1)^{n+1} \frac{\Gamma(n+\lambda)}{n!} e^{\pi i (n+\lambda)/2} x^{-n-\lambda} e^{ixa} \qquad n = 0, 1, \dots .$$

We can now apply 2.1 (5). The contributions of β to the s_n vanish since $\phi^{(n)}(\beta) = 0$. From (10) it follows that $\Sigma \, s_n = -A_N$, where A_N is given by (5). Moreover, from 2.1 (7),

$$R_N = (-1)^N \int_a^{\beta} \phi^{(N)}(t) \, h_{-N}(t) \, dt,$$

and by (9),

$$|R_N| \leq x^{-N} \int_a^{\beta} |\phi^{(N)}(t)| \, (t-a)^{\lambda-1} \, dt = O(x^{-N}).$$

This proves (4).

Lastly we turn to Fourier integrals whose integrands have singularities at both ends of the interval.

If $\phi(t)$ *is* N *times continuously differentiable for* $a \leq t \leq \beta$, *and* $0 < \lambda \leq 1$, $0 < \mu \leq 1$, *then*

$$(11) \quad \int_a^{\beta} e^{ixt} (t-a)^{\lambda-1} (\beta - t)^{\mu-1} \phi(t) \, dt = B_N(x) - A_N(x) + O(x^{-N})$$

as $x \to \infty$,

where

$$(12)\quad A_N(x) = \sum_{n=0}^{N-1} \frac{\Gamma(n+\lambda)}{n!} e^{\pi i(n+\lambda-2)/2} x^{-n-\lambda} e^{ix\alpha} \frac{d^n}{d\alpha^n}[(\beta-\alpha)^{\mu-1}\phi(\alpha)]$$

$$B_N(x) = \sum_{n=0}^{N-1} \frac{\Gamma(n+\mu)}{n!} e^{\pi i(n-\mu)/2} x^{-n-\mu} e^{ix\beta} \frac{d^n}{d\beta^n}[(\beta-\alpha)^{\lambda-1}\phi(\beta)],$$

and $O(x^{-N})$ *in* (11) *may be replaced by* $o(x^{-N})$ *if* $\lambda = \mu = 1$.

This theorem contains the three previous results as special cases. To prove (11), we shall use a device which is frequently employed for such purposes, and is called a *neutralizer* by van der Corput. Let $\nu(t)$ be an infinitely differentiable function for $\alpha \le t \le \beta$, $\nu(\alpha) = 1$, $\nu^{(n)}(\alpha) = 0$, $n = 1, 2, \dots$; $\nu^{(n)}(\beta) = 0$, $n = 0, 1, 2, \dots$. An example of such a function is

$$\frac{\int_t^\beta \exp\left(-\dfrac{1}{u-\alpha} - \dfrac{1}{\beta-u}\right) du}{\int_\alpha^\beta \exp\left(-\dfrac{1}{u-\alpha} - \dfrac{1}{\beta-u}\right) du}.$$

With such a neutralizer $\nu(t)$ we write

$$(13)\quad \int_\alpha^\beta e^{ixt}(t-\alpha)^{\lambda-1}(\beta-t)^{\mu-1}\phi(t)\,dt$$

$$= \int_\alpha^\beta e^{ixt}(t-\alpha)^{\lambda-1}[\nu(t)(\beta-t)^{\mu-1}\phi(t)]\,dt$$

$$+ \int_\alpha^\beta e^{ixt}(\beta-t)^{\mu-1}\{[1-\nu(t)](t-\alpha)^{\lambda-1}\phi(t)\}\,dt.$$

The first integral on the right hand side is of the form (4), with $\phi(t)$ replaced by [...]: since all derivatives of this function vanish at $t = \beta$, and are equal to the corresponding derivatives of $(\beta - t)^{\mu-1}\phi(t)$ at $t = \alpha$, we obtain the expression (12) for $A_N(x)$. Similarly, the second integral on the right-hand side of (13) is of the form (6), with $\phi(t)$ replaced by { ... }; all derivatives of { ... } vanish at $t = \alpha$, and are equal to the corresponding derivatives of $(t-\alpha)^{\lambda-1}\phi(t)$ at $t = \beta$; and the expression (12) for $B_N(x)$ follows from (7). This proves (11). If $\lambda = \mu = 1$, then (2) shows that $O(x^{-N})$ may be replaced by $o(x^{-N})$.

All our results remain true if i is replaced by $-i$ throughout the formulas, and in (8) the integration from t to $-i\infty$ is taken over a path which lies in the quadrant $-\pi/2 \le \arg(u-a) \le 0$. Thus we are able to describe the behavior of our integrals as $x \to -\infty$, and also the asymptotic behavior of Fourier integrals with trigonometric kernels $\cos xt$ and $\sin xt$.

2.9. The method of stationary phase

We now consider the integral

$$(1) \quad f(x) = \int_a^\beta g(t)\, e^{ixh(t)}\, dt$$

in which x is a large positive variable and $h(t)$ is a real function of the real variable t. According to Stokes and Kelvin, the major contribution to the value of the integral arises from the immediate vicinity of the end points of the interval and from the vicinity of those points at which $h(t)$ is *stationary*, i.e., $h'(t) = 0$; and in the first approximation the contribution of stationary points, if there are any, is more important then the contribution of the end points.

Suppose that g is continuous and h is twice continuously differentiable, let τ be the only stationary point of h, $a < \tau < \beta$, $h'(\tau) = 0$ and $h''(\tau) > 0$. In the assumption that the neighborhood of τ will give the principal contribution to the integral, we introduce a new variable of integration u by the substitution $h(t) - h(\tau) = u^2$ and obtain

$$f(x) \sim \int_{\tau-\epsilon}^{\tau+\epsilon} g(t)\, e^{ixh(t)}\, dt = \int_{-u_1}^{u_2} 2u \frac{g(t)}{h'(t)} \exp\{ix[h(\tau) + u^2]\}\, du$$

where $u_1 = [h(\tau-\epsilon) - h(\tau)]^{1/2}$, $u_2 = [h(\tau+\epsilon) - h(\tau)]^{1/2}$. Since only the neighborhood of $u = 0$ matters, we may replace $g(t)$ by $g(\tau)$ and $2u/h'(t)$ by $[2/h''(\tau)]^{1/2}$ which is the limit of $2u/h'(t)$ as $t \to \tau$, so that

$$f(x) \sim \left[\frac{2}{h''(\tau)}\right]^{1/2} g(\tau) \int_{-u_1}^{u_2} \exp[ixu^2 + ixh(\tau)]\, du \,.$$

By the same argument we may extend the integration from $-\infty$ to ∞ and finally obtain

$$(2) \quad f(x) \sim \left[\frac{2\pi}{xh''(\tau)}\right]^{1/2} g(\tau) \exp[ixh(\tau) + i\pi/4] \qquad \text{as } x \to \infty,$$

which is virtually Kelvin's result. The contribution of the point of stationary phase, τ, to the integral is more important than the contributions of the end points because the latter can be shown, by integration by parts, to be $O(x^{-1})$ if $h'(\alpha) \neq 0$, $h'(\beta) \neq 0$.

The principle of stationary phase has been applied to numerous mathematical and physical problems but it appears to be difficult to formulate it in a precise manner. Perhaps the best available theorem is one given by Watson (1920). Poincaré discussed the principle of stationary phase applied to integrals involving analytic functions, and the connection of his work with the method of steepest descents is indicated in Copson (1946). The method of stationary phase has also been discussed by Bijl (1937), and in a much more general setting by van der Corput (1934, 1936).

We shall use the discussion of Fourier integrals given in the preceding section to derive a theorem which may be regarded as a precise version, and at the same time generalization, of (2). A point τ at which $h'(\tau) = h''(\tau) = \dots = h^{(m)}(\tau) = 0$ and $h^{(m+1)}(\tau) \neq 0$ will be called a *stationary point of order* m, $m = 1, 2, \dots$. In the neighborhood of such a point $h'(t) = (t - \tau)^m h_1(t)$, where $h_1(\tau) \neq 0$. The notion of a stationary point may be generalized to fractional order. A point τ will be called a *stationary point of (fractional) order* μ if in some neighborhood of that point $h'(t)$ is either of the form $|t - \tau|^{\mu} h_1(t)$ or of the form $\operatorname{sgn}(t - \tau)\,|t-\tau|^{\mu} h_1(t)$, where $h_1(\tau) \neq 0$. Assuming that $h(t)$ has at most a finite number of stationary points (of positive order) in the interval under consideration, we may break up the integral in a finite number of integrals in each of which $h(t)$ is monotonic; and we may assume $h(t)$ to be *increasing*. Thus, we shall consider integrals of the form (1) in which $h(t)$ is strictly increasing when $\alpha < t < \beta$, and α and β are either ordinary points (i.e., stationary points of order zero), or stationary points (of positive order).

If $0 < \lambda$, $\mu \leq 1$; $g(t)$ *is* N *times continuously differentiable for* $\alpha \leq t \leq \beta$; $h(t)$ *is differentiable and*

$$(3) \quad h'(t) = (t - \alpha)^{\rho-1} (\beta - t)^{\sigma-1} h_1(t),$$

where $\rho, \sigma \geq 1$, *and* $h_1(t)$ *is positive and* N *times continuously differentiable for* $\alpha \leq t \leq \beta$; *then*

$$(4) \quad \int_{\alpha}^{\beta} g(t)\, e^{ixh(t)} (t - \alpha)^{\lambda-1} (\beta - t)^{\mu-1}\, dt = B(x) - A(x)$$

where

(5) $A(x) \sim A_N(x)$ and $B(x) \sim B_N(x)$ to N terms as $x \to \infty$,

and $A_N(x)$ *and* $B_N(x)$ *are given by* (17) and (20) *below.*

In the proof of this theorem we shall use the abbreviation

(6) $g_1(t) = g(t)\,(t-a)^{\lambda-1}\,(\beta-t)^{\mu-1}$,

and shall employ a *neutralizer* $\nu(t)$, which is infinitely differentiable for $a \leq t \leq \beta$ and such that for some η, $0 < \eta < (\beta-a)/2$, $\nu(t) = 1$ when $a \leq t \leq a+\eta$, and $\nu(t) = 0$ when $\beta - \eta \leq t \leq \beta$. We then set

$$(7)\quad -A(x) = \int_a^{\beta-\eta} \nu(t)\, g_1(t)\, e^{ixh(t)}\, dt$$

$$(8)\quad B(x) = \int_{a+\eta}^{\beta} [1-\nu(t)]\, g_1(t)\, e^{ixh(t)}\, dt.$$

To obtain an asymptotic expansion of $A(x)$, we introduce a new variable of integration, u, in (7) by

$$(9)\quad u^\rho = h(t) - h(a), \qquad u_1^\rho = h(\beta-\eta) - h(a).$$

From (3) we have

$$u^\rho = h(t) - h(a) = \int_a^t h'(s)\, ds$$

$$= (t-a)^\rho \int_0^1 y^{\rho-1}[\beta - a - (t-a)y]^{\sigma-1}\, h_1[a + (t-a)\,y]\, dy$$

where $s = a + (t-a)y$. The last integral is an $N+1$ times continuously differentiable, positive, and increasing function of t, so that (9) represents an $N+1$ times continuously differentiable mapping of the interval $a \leq t \leq \beta - \eta$ onto the interval $0 \leq u \leq u_1$, and the inverse mapping is also $N+1$ times continuously differentiable.

We now put $\nu_1(u) = \nu(t)$ and

$$(10)\quad k(u) = g_1(t)\, u^{1-\lambda}\, \frac{dt}{du},$$

where $g_1(t)$ is given by (6), and $k(u)$ is N times continuously differentiable for $0 \le u \le u_1$. Then

$$A(x) = -e^{ixh(\alpha)} \int_0^{u_1} \nu_1(u)\, k(u)\, u^{\lambda-1} \exp(ixu^{\rho})\, du$$

can be integrated by parts N times, differentiating $\nu_1 k$ and integrating the remaining factor of the integrand. With

$$(11)\quad \phi_{-n-1}(u) = \frac{(-1)^{n+1}}{n!} \int_u^{\infty} (z-u)^n z^{\lambda-1} \exp(ixz^{\rho})\, dz$$

the result of the integrations by parts is

$$A(x) = A_N(x) + R_N(x),$$

where

$$(12)\quad A_N(x) = \sum_{n=0}^{N-1} (-1)^n k^{(n)}(0)\, \phi_{-n-1}(0)\, e^{ixh(\alpha)}$$

and

$$(13)\quad R_N(x) = (-1)^{N+1} e^{ixh(\alpha)} \int_0^{u_1} \phi_{-N}(u) \frac{d^N(\nu_1 k)}{du^N}\, du.$$

In (11), the path of integration is the ray $\arg(z-u) = \pi/(2\rho)$ in the complex plane. Clearly,

$$(14)\quad \phi_{-n-1}(0) = \frac{(-1)^{n+1}}{n!\,\rho} \Gamma\left(\frac{n+\lambda}{\rho}\right) \exp\left[\frac{\pi i(n+\lambda)}{2\rho}\right] x^{-(n+\lambda)/\rho}.$$

To estimate $\phi_{-n-1}(u)$ for $u > 0$, we note that $|z|^{\lambda-1} \le u^{\lambda-1}$ and also that

$$ixz^{\rho} + x|z-u|^{\rho} = i\rho x \int_0^{u} \left(\xi + |z-u| \exp\frac{\pi i}{2\rho}\right)^{\rho-1} d\xi.$$

Since the real part of the last expression is certainly negative, we have

$$|\exp(ixz^{\rho})| \le \exp(-x|z-u|^{\rho})$$

and hence

$$(15)\quad |\phi_{-n-1}(u)| \le \frac{u^{\lambda-1}}{n!}\int_u^\infty |z-u|^n \exp(-x|z-u|^\rho)\, d|z-u|$$

$$\le \frac{1}{n!}\Gamma\left(\frac{n+1}{\rho}\right) u^{\lambda-1} x^{-(n+1)/\rho}.$$

Alternatively, the method of steepest descents may be applied to (11) to show that

$$(16)\quad \phi_{-n-1}(u) = u^{n+\lambda} O[(xu^\rho)^{-n-1}]$$

for large xu^ρ.

Substituting (14) and (15) in (12) and (13) we obtain

$$(17)\quad A_n(x) = -\sum_{n=0}^{N-1} \frac{k^{(n)}(0)}{n!\,\rho}\Gamma\left(\frac{n+\lambda}{\rho}\right)\exp\left[\frac{\pi i(n+\lambda)}{2\rho}\right] x^{-(n+\lambda)/\rho} e^{ixh(\alpha)}$$

$$|R_N(x)| \le \frac{1}{(N-1)!}\Gamma\left(\frac{N}{\rho}\right) x^{-N/\rho}\int_0^{u_1} u^{\lambda-1}\left|\frac{d^N(\nu_1 k)}{du^N}\right| du.$$

This proves that $A \sim A_N$ to N terms when $\lambda < 1$. When $\lambda = 1$ and $\rho = 1$ the same result follows from sec. 2.8. Let $\lambda = 1$ and $\rho > 1$ and choose δ so that

$$\frac{1}{(N-1)!}\Gamma\left(\frac{N}{\rho}\right)\int_0^\delta u^{\lambda-1}\left|\frac{d^N(\nu_1 k)}{du^N}\right| du < \tfrac{1}{2}\epsilon.$$

Since (16) gives $\phi_{-N}(u) = O(x^{-N})$ uniformly in u when $u \ge \delta$, we have for sufficiently large x

$$x^{N/\rho}\int_\delta^{u_1} |\phi_{-N}(u)| \left|\frac{d^N(\nu_1 k)}{du^N}\right| du < \tfrac{1}{2}\epsilon$$

so that $R_N = o(x^{-N/\rho})$ also in this case. This proves the result for A.

A similar result holds for $B(x)$. In (8) we introduce a new variable of integration by

(18) $v^{\sigma} = h(\beta) - h(t)$

and put

(19) $l(v) = g_1(t)\, v^{1-\mu} \dfrac{dt}{dv},$

where $g_1(t)$ is given by (6). In the repeated integrals of $v^{\mu-1} \exp(-ixv^{\sigma})$ we integrate along the ray $\arg(z - v) = -\pi i/(2\sigma)$, and obtain by a process very similar to that used in the case of A that $B \sim B_N$ to N terms where

(20) $$B_N(x) = -\sum_{n=0}^{N-1} \frac{l^{(n)}(0)}{n!\,\sigma}\, \Gamma\left[\frac{n+\mu}{\sigma}\right] \exp\left[\frac{-\pi i(n+\mu)}{2\sigma}\right] \times x^{-(n+\mu)/\sigma}\, e^{ixh(\beta)}.$$

We conclude this section by applying the general result to

$$f(x) = \int_0^1 \exp(ixt^3)\, dt.$$

Here $\lambda = \mu = 1,\ \rho = 3,\ \sigma = 1,\ u = t,\ k(u) = 1$ and

$$A_N(x) = 1/3\, \Gamma(1/3)\, e^{\pi i/6}\, x^{-1/3}$$

by (17). Also $v = 1 - t^3$, $t = (1-v)^{1/3}$,

$$l(v) = \frac{dt}{dv} = -\frac{1}{3}(1-v)^{-2/3},$$

and

$$B_N(x) = -\sum_{n=0}^{N-1} \frac{\Gamma(n+2/3)}{\Gamma(-1/3)} (ix)^{-n-1} e^{ix}$$

so that

$$f(x) \sim \Gamma(4/3)\, e^{\pi i/6}\, x^{-1/3} - \sum_{n=0}^{\infty} \frac{\Gamma(n+2/3)}{\Gamma(-1/3)} (ix)^{-n-1} e^{ix} \qquad \text{as } x \to \infty,$$

which is the expansion obtained in sec. 2.7 by the method of steepest descents. Note that $x \to \infty$ through positive values with our present method while x could be complex in sec. 2.7.

REFERENCES

Bijl, Jan, 1937: *Toepassingen van der methode der stationnaire phase* (Thesis). Amsterdam.

Burkhardt, Heinrich, 1914: *München. Akad. S. Ber.* 1-11.

Copson, E.T., 1935: *An introduction to the theory of functions of a complex variable*. Oxford

Copson, E.T., 1946: *The asymptotic expansion of a function defined by a definite integral or a contour integral.* Admiralty Computing Service. London.

van der Corput, J.G., 1934: *Composito Math.* 1, 15-38.

van der Corput, J.G., 1936: *Compositio Math.* 3, 328-372.

van der Corput, J.G., 1948: *Proc. Amst. Akad. Wet.* 51, 650-658.

van der Corput, J.G., and Franklin, Joel, 1951: *Proc. Amsterdam Akad. A.* 54, 213-219.

Doetsch, Gustav, 1950: *Handbuch der Laplace-Transformation.* Birkhäuser, Basel.

Erdélyi, Arthur, 1947: *Proc. Edinburgh Math. Soc.* (2) 8, 20-24.

Fulks, W.B., 1951: *Proc. Amer. Math. Soc.* 2, 613-622.

Hsu, L.C., 1948a: *Amer. J. Math.* 70, 698-708.

Hsu, L.C., 1948b: *Duke Math. J.* 15, 623-632.

Hsu, L.C., 1949a: *Sci. Rep. Nat. Tsing Hua Univ. Ser. A* 5, 273-279.

Hsu, L.C., 1949b: *Acad. Sinica Science Record* 2, 339-345.

Hsu, L.C., 1951a: *Chung Kuo K'o Hsueh* (*Chinese Science*) 2, 149-155.

Hsu, L.C., 1951b: *Bull. Calcutta Math. Soc.* 43, 109-112.

Hsu, L.C., 1951c: *Amer. J. Math.* 73, 625-634.

Levi, Beppo, 1946: *Publ. Inst. Mat. Univ. Nac. Litoral* 6, 341-351.

Meijer, C.S., 1933a: *Math. Ann.* 108, 321-359.

Meijer, C.S., 1933b: *Asymptotische Entwickelungen Besselscher, Hankelscher und verwandter Funktionen.* Groningen (Thesis).

Perron, Oskar, 1917: *München. Akad. S. Ber.* 191-220.

Rooney, P.G., 1953: *Trans. Roy. Soc. Canada, Sect. III.* 47, 29-34.

Thomsen, D.L., Jr., 1954: *Proc. Amer. Math. Soc.* 5, 526-532.

Watson, G.N., 1920: *Proc. Cambridge Philos. Soc.* 19, 49-55.

Widder, D.V., 1941: *The Laplace transform.* Princeton.

CHAPTER III

SINGULARITIES OF DIFFERENTIAL EQUATIONS

In this chapter we give a brief introduction to the asymptotic theory of *ordinary homogeneous linear differential equations of the second order.* Analogous theories exist for equations of arbitrary (finite) order and for systems of differential equations of the first order. For these more general theories see Ince (1927, especially p. 169 ff., p. 428 ff., p. 444 ff., p. 484 ff.), Kamke (1944, especially, p. 17 ff., p. 60 ff., p. 100 ff., p. 133 ff.), Wasow (1953), the references given in these works, and the references given at the end of the present chapter. Asymptotic expansions occur also in connection with non-linear differential equations, and partial differential equations.

We shall investigate the asymptotic behavior of solutions of

$$y'' + p(x)\,y' + q(x)\,y = 0$$

as $x \to x_0$. Here x is either a real variable ranging over an interval (of which x_0 is usually an end-point), or else a complex variable ranging over a region (of which x_0 is often a boundary point). Without loss of generality, we take $x_0 = \infty$ throughout this chapter.

The reader is expected to know the basic existence theorems regarding the above differential equation both for real and complex variables, and he is also expected to be familiar with the fundamental properties of the solutions.

3.1. Classification of singularities

In the present section we shall discuss the differential equation

$$(1) \quad y'' + p(x)\,y' + q(x)\,y = 0$$

when x is a complex variable ranging over an annular region, R, given by $r < |x| < \infty$, and $p(x)$ and $q(x)$ are single-valued analytic functions in R

(which may or may not have singularities at ∞). We shall briefly review the well-known classification (see, for instance, Poole, 1936, Chapter IV) of isolated singularities of (1), the singularity in question being at ∞.

If $y_1(x)$ and $y_2(x)$ are two linearly independent solutions of (1) forming a fundamental system, and if we continue these functions analytically along some curve in R which begins and ends at x and encircles ∞ in the positive sense, we obtain two new functions which may be denoted by $y_j(xe^{-2\pi i})$, $j = 1, 2$. These need not be identical with the $y_j(x)$, but at any rate, they will be solutions of (1), so that relations of the form

$$
\begin{aligned}
&y_1(xe^{-2\pi i}) = a_{11}\, y_1(x) + a_{12}\, y_2(x) \\
(2)\qquad & \\
&y_2(xe^{-2\pi i}) = a_{21}\, y_1(x) + a_{22}\, y_2(x)
\end{aligned}
$$

will hold. In (2),

$$
A = \begin{bmatrix} a_{11} & a_{12} \\ a_{21} & a_{22} \end{bmatrix}
$$

is a constant non-singular matrix.

If instead of y_1, y_2 we take another fundamental system, we obtain a matrix B, and an easy computation shows that $B = MAM^{-1}$, where M is a non-singular constant matrix. Thus, all matrices obtained in this manner have the same affine *invariants*, in particular, the same *latent roots*, and the same *canonical forms*. These, then, are independent of the fundamental system chosen, and are characteristic of the singularity at ∞ (if there is a singularity there).

Let us assume that the latent roots of A are *distinct* so that we have a *diagonal canonical form*

$$
\begin{bmatrix} \lambda_1 & 0 \\ 0 & \lambda_2 \end{bmatrix}.
$$

If y_1, y_2 is the fundamental system corresponding to the canonical form of A (*canonical fundamental system*), then (2) assumes the form

$$
(3)\qquad y_j(xe^{-2\pi i}) = \lambda_j\, y_j(x) \qquad j = 1, 2.
$$

We now set $\lambda_j = \exp(2\pi i \rho_j)$, and call ρ_1, ρ_2 the *exponents* belonging to ∞: these are determined up to an integer in each. From (3) we see that the canonical fundamental system is of the form

$$(4) \quad y_j(x) = x^{-\rho_j} \psi_j(x) \qquad j = 1,2$$

where ψ_1, ψ_2 are single-valued analytic functions of x in R with, possibly, singularities at ∞.

If the latent roots are equal, $\lambda_1 = \lambda_2 = \lambda = \exp(2\pi i \rho)$, then the canonical fundamental system can be shown to be of the form

$$(5) \quad \begin{aligned} y_1(x) &= x^{-\rho} \psi_1(x) \\ y_2(x) &= c y_1(x) \log x + x^{-\rho} \psi_2(x), \end{aligned}$$

where ψ_1 and ψ_2 have the same properties as in (4), and c is a constant: $c = 0$ or $c \neq 0$ according as the canonical form of A in this case is diagonal or not.

$x = \infty$ is called an *ordinary point* of (1) if all solutions are regular at ∞, i.e., can be represented by convergent power series in x^{-1}; $x = \infty$ is called a *regular singularity* of (1) if it is not an ordinary point and if ψ_1 and ψ_2 have at most poles at ∞ so that by a suitable choice of ρ_1 and ρ_2, ψ_1 and ψ_2 can be made regular at ∞; and $x = \infty$ is called an *irregular singularity* of (1) if at least one of the two functions ψ_1, ψ_2 has an essential singularity at ∞.

It can be shown (see, for instance, Poole, 1936, section 20) that a *sufficient condition* for $x = \infty$ to be an *ordinary point* is

$$(6) \quad p(x) = 2x^{-1} + O(x^{-2}), \qquad q(x) = O(x^{-4}) \qquad \text{as } x \to \infty,$$

and that a *sufficient condition* for $x = \infty$ to be a *regular singularity* is

$$(7) \quad p(x) = O(x^{-1}), \qquad q(x) = O(x^{-2}) \qquad \text{as } x \to \infty.$$

In the case of an *irregular singularity*, p and q may have essential singularities at ∞: if p and q have at most poles at ∞, we speak of an irregular singularity of finite rank, and the least integer k for which

$$(8) \quad p(x) = O(x^{k-1}), \qquad q(x) = O(x^{2k-2}) \qquad \text{as } x \to \infty$$

is called the *rank* of the irregular singularity. Sometimes, a regular singularity is regarded as a singularity of rank zero.

3.2. Normal solutions

If $x = \infty$ is an ordinary point of 3.1(1), we may expand y in a series of powers of x^{-1}. The coefficients of that series may be determined from recurrence relations, and the series converges in R. If $x = \infty$ is a regular singularity of 3.1(1), we may set

$$y = \sum_{n=0}^{\infty} c_n x^{-\rho-n}, \qquad c_0 \neq 0.$$

We then obtain a quadratic equation for ρ, recurrence relations for the c_n (similar to (7) below), and a series for y which converges in R. In either of these two cases the coefficients can be computed easily, and the convergent series can be used with advantage to compute the solutions for large x.

The situation is entirely different if $x = \infty$ is an irregular singular point. Since the ψ_j have essential singularities at ∞, we must set in this case

$$y = \sum_{n=-\infty}^{\infty} c_n x^{-\rho-n}.$$

For the c_n we obtain an infinite system of linear equations which cannot be solved recurrently, and for ρ, a transcendental equation which involves an infinite determinant (the determinant of the system). In this case the coefficients cannot be computed easily, nor is the series rapidly convergent for large x.

It was discovered by Thomé that in the case of an *irregular singularity of finite rank* certain formal solutions exist which do not suffer from the disadvantages mentioned above; the coefficients occurring in these solutions can be computed recurrently, and the series appear suitable for numerical computations for large x. Thomé's solutions are of the form

$$y = \exp\,[P(x)] \sum_{n=0}^{\infty} c_n x^{-\rho-n}, \qquad c_0 \neq 0,$$

where $P(x)$ is a polynomial: they are known as *normal solutions*.

We shall explain the construction of normal solutions in the case of an *irregular singularity of rank one*. We first note that setting

$$y = z \exp(-\tfrac{1}{2} \textstyle\int p\, dx)$$

in 3.1(1), we obtain for z a differential equation of the form (1) in which z' does not occur. Hence it will be sufficient to discuss the differential equation

(1) $y'' + q(x)y = 0$

in which

(2) $q(x) = \sum_{n=0}^{\infty} q_n x^{-n},$

the series being convergent in R. We shall attempt to find formal solutions of the form

(3) $y = e^{\omega x} \sum_{n=0}^{\infty} c_n x^{-\rho-n}, \qquad c_0 \neq 0,$

where ω and ρ are constants. It is to be noted that ρ in (3) is not necessarily one of the exponents belonging (in the sense of the analytical theory of sec. 3.1) to the irregular singularity $x = \infty$.

In manipulating the formal series, we shall adopt the convention $q_{-m} = 0$, $c_{-m} = 0$, $m = 1, 2, \ldots$, so that all summations may be extended from $-\infty$ to $+\infty$ except those explicitly stated otherwise.

Substituting (2) and (3) in (1), we have

$$\omega^2 \sum c_n x^{-\rho-n} - 2\omega \sum (\rho+n) c_n x^{-\rho-n-1}$$
$$+ \sum (\rho+n)(\rho+n+1) c_n x^{-\rho-n-2} + \sum q_n x^{-n} \sum c_n x^{-\rho-n} = 0.$$

Comparing coefficients here, we obtain

(4) $$\omega^2 c_n - 2\omega(\rho+n-1) c_{n-1} + (\rho+n-2)(\rho+n-1) c_{n-2} + \sum_{\nu=0}^{n} q_\nu c_{n-\nu} = 0$$

for all integers n. The first non-vacuous condition arises when $n = 0$ in (4). Since $c_0 \neq 0$ we have

(5) $\omega^2 + q_0 = 0.$

If $n = 1$ in (4), and ω satisfies (5), we obtain

$$(6) \quad -2\omega\rho + q_1 = 0.$$

These two equations determine ω and ρ. The recurrence relation for the coefficients may also be obtained from (4). We replace n by $n+1$, and use (5) and (6) to obtain

$$(7) \quad 2\omega n c_n = (\rho+n)(\rho+n-1)\, c_{n-1} + \sum_{\nu=2}^{n+1} q_\nu c_{n+1-\nu} \qquad n = 1, 2, \ldots .$$

We now see that normal solutions exist if either $q_0 \neq 0$, or $q_0 = q_1 = 0$. In the former case, (5) determines ω, (6) determines ρ, and with $c_0 = 1$, (7) determines the coefficients. Moreover, $\omega, \rho, c_1, \ldots, c_m$ are completely determined by $q_0, q_1, \ldots, q_{m+1}$, and vice versa. There are in this case always two normal solutions corresponding to the two possible values of ω. In the latter case, which is the case of a *regular singularity* at ∞, we have $\omega = 0$ from (5), equation (6) is vacuous, (7) with $n = 1$ determines ρ as one of the roots of the quadratic equation $\rho(\rho+1) + q_2 = 0$, and (7) with $n = 2, 3, \ldots$ determines the coefficients.

If $q_0 = 0$ and $q_1 \neq 0$, then (5) and (6) cannot be satisfied, and there exists no normal solution. However, *subnormal solutions* may be obtained by transforming (1) into

$$(8) \quad \eta'' + \left[4\xi^2 q(\xi^2) - \frac{3}{4\xi^2} \right] \eta = 0$$

by the change of variables

$$\xi = x^{1/2}, \qquad \eta(\xi) = \xi^{-1/2}\, y(x).$$

If $q_0 = 0$ and $q_1 \neq 0$, then (8) has an irregular singularity of rank one at ∞, and possesses normal solutions. These give rise to subnormal solutions of (1) having the form

$$(9) \quad y = \exp(\omega x^{1/2}) \sum_{n=0}^{\infty} c_n x^{1/4 - 1/2 n}$$

in this case.

For the construction of normal and subnormal solutions for singularities of higher ranks, for differential equations of higher orders, and for systems of differential equations see Ince (1927, p. 423 ff., p. 427 ff., p. 469 ff., p. 478 ff.).

Normal and subnormal solutions are *formal solutions*; i.e., if they are substituted in the differential equation as if the infinite series were convergent, the differential equation is satisfied. However, the infinite series involved in the formal solutions are in general *divergent*. Nevertheless, they are far from being useless, for they represent *asymptotic expansions* of solutions of (1). This situation has been investigated by many authors, beginning with Poincaré: some of the papers are listed at the end of this chapter, and it may be noted that the most general results were obtained by Sternberg (1920) for the differential equation of order n and arbitrary (finite) rank, and by Trjitzinsky (1933) for a system of first order differential equations.

By and large there seem to be two methods for proving that the formal solutions are asymptotic expansions of solutions of the differential equation. One of these was originated by Poincaré and was developed by Horn in a large number of papers of which we refer but to a few at the end of this chapter. This method consists in finding integral representations of Laplace's type for the solutions, and then basing the asymptotic expansions on the work of sec. 2.2. The other method, which was developed by G.D. Birkhoff and his pupils, uses the leading terms, or partial sums, of the formal solutions to construct a differential equation which in a certain sense is close to the given equation when x is large, and then compares the two equations. As it happens, singular Volterra integral equations or integro-differential equations play an important part in both methods.

We shall use a variant of the second method to discuss the differential equation (1) with $q_0 \neq 0$. The proof to be given below is based principally on the work of Hoheisel (1924) and Tricomi (1953, sections 47 to 50). Since the analytic character of q and y does not enter in the formal solution, the investigation may be carried out either for real or for complex independent variables.

3.3. The integral equation and its solution

We first consider the case of a real variable, and defer a brief discussion of the case of a complex variable to sec. 3.5. Let us then consider

$$(1) \quad y'' + q(x)\, y = 0 \qquad x \geq a > 0,$$

assuming that $q(x)$ is continuous for $x \geq a$ and

$$(2) \quad q(x) \sim \sum_{n=0}^{\infty} q_n x^{-n} \qquad \text{as } x \to \infty,\ q_0 \neq 0.$$

We then obtain two formal solutions

$$(3) \quad e^{\omega x} \sum_{n=0}^{\infty} c_n x^{-\rho-n}, \qquad c_0 \neq 0$$

where ω, ρ, c_1, c_2, ... satisfy 3.2(5), (6), (7). We shall show that *these formal solutions are asymptotic expansions of certain solutions of* (1).

Let $\omega = \omega_1 + i\omega_2$, $\rho = \rho_1 + i\rho_2$, and determine $\omega = (-q_0)^{1/2}$ so as to make $e^{\omega x} x^{-\rho}$ bounded as $x \to \infty$. If q_0 is not positive real, we take that value of the square root which makes $\omega_1 < 0$; if $q_0 > 0$ and $\operatorname{Im} q_1 \neq 0$, we take that value of the square root which makes $\rho_1 = \operatorname{Re}[q_1/(2\omega)] > 0$; and if $q_0 > 0$ and q_1 is real, we take either value of the square root. Thus, we always have *either* $\omega_1 < 0$ *or* $\omega_1 = 0$ and $\rho_1 \geq 0$. These conventions will be retained throughout the discussion.

It will be convenient to transform (1) by setting

$$(4) \quad y(x) = e^{\omega x} x^{-\rho} z(x)$$

so that z satisfies the differential equation

$$(5) \quad z'' + 2\left(\omega - \frac{\rho}{x}\right) z' + \left[\omega^2 - \frac{2\omega\rho}{x} + \frac{\rho(\rho+1)}{x^2} + q(x)\right] z = 0.$$

Here ω and ρ satisfy 3.2(5), (6). We put

$$(6) \quad x^2[q(x) - q_0 - q_1 x^{-1}] + \rho(\rho+1) = F(x),$$

and see from (2) that $F(x)$ is bounded, say

$$(7) \quad |F(x)| \leq A, \qquad x \geq a.$$

We now rewrite (5) as

$$\frac{d}{dx}\left(e^{2\omega x} x^{-2\rho} \frac{dz}{dx}\right) + e^{2\omega x} x^{-2\rho-2} F(x)\, z(x) = 0,$$

integrate to obtain

$$\frac{dz}{dx} + e^{-2\omega x} x^{2\rho} \int_b^x e^{2\omega t} t^{-2\rho-2} F(t)\, z(t)\, dt = c_2\, x^{-2\omega x} x^{2\rho},$$

where c_2 and $b \geq a$ are arbitrary constants, and integrate once more to obtain

$$(8) \quad z(x) + \int_b^x K(x, t)\, F(t)\, z(t)\, t^{-2}\, dt = c_1 + c_2 \int_a^x e^{-2\omega t} t^{2\rho}\, dt,$$

where

$$(9) \quad K(x, t) = -\int_x^t \exp[2\omega(t-s)] \left(\frac{s}{t}\right)^{2\rho} ds.$$

Equation (8) is an integral equation of Volterra's type. Any solution of (5) satisfies (8) for some b, c_1, c_2, and conversely, any twice continuously differentiable solution of (8), for any b, c_1, c_2, satisfies (5). The existence of such a solution follows from the general theory of integral equations when $b < \infty$. When $b = \infty$, the integral equation (8) is a singular integral equation, and the existence and differentiability of the solution must be demonstrated.

In order to prove that (5) possesses a solution which can be represented asymptotically by $\Sigma\, c_n x^{-n}$, we take $b = \infty$, $c_1 = 1$, $c_2 = 0$ in (8) so that the integral equation (8) becomes

$$(10) \quad z(x) = 1 + \int_x^\infty K(x, t)\, F(t)\, z(t)\, t^{-2}\, dt.$$

This integral equation will be solved by the method of successive approximations.

For any function, $\zeta(x)$, we set

$$(11) \quad T\zeta(x) = \int_x^\infty K(x, t)\, F(t)\, \zeta(t)\, t^{-2}\, dt,$$

and then define

$$(12) \quad z_0(x) = 1, \qquad z_{n+1}(x) = Tz_n(x) \qquad\qquad n = 0, 1, 2, \ldots$$

$$(13) \quad z(x) = \sum_{n=0}^{\infty} z_n(x).$$

It will now be proved that $z(x)$ exists, satisfies (10), is differentiable, and satisfies (5). The proof will be conducted in several steps.

The kernel, $K(x, t)$, is bounded for $t \geq x \geq x_0$ where $x_0 > a$ and x_0 is sufficiently large.

Proof: Since either $\omega_1 < 0$ or $\omega_1 = 0$ and $\rho_1 \geq 0$ we have

$$\frac{d}{ds} \log \left(e^{-2\omega_1 s} s^{2\rho_1}\right) = -2\omega_1 + \frac{2\rho_1}{s} \geq 0$$

for sufficiently large s; and hence

$$e^{-2\omega_1 s} s^{2\rho_1}$$

is an increasing function of s. We now write

$$e^{2\omega(t-s)} \left(\frac{s}{t}\right)^{2\rho} = e^{2\omega_1(t-s)} \left(\frac{s}{t}\right)^{2\rho_1} [\phi_1(s, t) + i\phi_2(s, t)]$$

where

$$\phi_1(s, t) + i\phi_2(s, t) = e^{2i\omega_2(t-s)} \left(\frac{s}{t}\right)^{2i\rho_2},$$

and apply the second mean value theorem to (9), obtaining

$$-K(x, t) = \int_x^t e^{2\omega_1(t-s)} \left(\frac{s}{t}\right)^{2\rho_1} [\phi_1(s, t) + i\phi_2(s, t)]\, ds$$

$$= e^{2\omega_1(t-x)} \left(\frac{x}{t}\right)^{2\rho_1} \left[\int_x^\xi \phi_1(s, t)\, ds + i \int_x^\eta \phi_2(s, t) ds\right]$$

$$+ \int_\xi^t \phi_1(s, t)\, ds + i \int_\eta^t \phi_2(s, t)\, ds,$$

where $x \leq \xi,\ \eta \leq t$. The integrals on the right-hand side are bounded functions of x and t, and

$$e^{2\omega_1(t-x)} \left(\frac{x}{t}\right)^{2\rho_1} \leq 1,$$

when $t \geq x \geq x_0$ and x_0 is sufficiently large, so that

(14) $|K(x, t)| \leq B, \quad t \geq x \geq x_0$

for some x_0 and B.

If $|\zeta(t)| \leq Ct^{-\lambda}$ *for* $t \geq x_0$, *where* $\lambda > -1$, *then*

(15) $$|T\zeta(x)| \leq \frac{ABC}{\lambda + 1} x^{-\lambda-1}, \qquad x \geq x_0.$$

We have by (11), (7), and (14)

$$|T\zeta(x)| = |\int_x^\infty K(x, t) F(t) \zeta(t) t^{-2} dt| \leq ABC \int_x^\infty t^{-\lambda-2} dt,$$

and this proves (15).

For the functions defined by (12) *we have*

(16) $$|z_n(x)| \leq \frac{(AB)^n}{n!} x^{-n}, \qquad x \geq x_0.$$

Proof by induction. (16) is true for $n = 0$, and if it is true for any n, the definition of z_{n+1} combined with (15) shows that it is also true for $n + 1$.

The series (13) *converges uniformly for* $x > x_0$, *and the function* $z(x)$ *satisfies* (10). *Moreover,* $z(x)$ *is twice continuously differentiable and satisfies* (5). The uniform convergence follows from (16). If we substitute $z = \Sigma z_n$ in the integral in (10), term-by-term integration is justified by uniform convergence, and shows that (10) is satisfied. Furthermore, the integral in (10) is a differentiable function of x, and so is $z(x)$. Since

$$K(x, x) = 0, \qquad \frac{\partial K(x, t)}{\partial x} = e^{2\omega(t-x)} \left(\frac{x}{t}\right)^{2\rho},$$

we obtain from (10),

(17) $$z'(x) = \int_x^\infty e^{2\omega(t-x)} \left(\frac{x}{t}\right)^{2\rho} F(t) z(t) t^{-2} dt.$$

The last integral is again a differentiable function of x, and substitution shows that $z(x)$ satisfies (5). With z given by (13),

(18) $y_1(x) = e^{\omega x} x^{-\rho} z(x)$

satisfies (1). If q_0 is positive and q_1 is real, we may take either of the two values of $(-q_0)^{1/2}$ for ω, and we thus obtain two linearly independent solutions of the form (18). In every other case, there is only one solution of this form, and a second solution may be written down in the form

(19) $y_2(x) = y_1(x) \int_b^x [y_1(t)]^{-2}\, dt,$

where b is any number large enough to ensure that $y_1(x) \neq 0$ for $x \geq b$. Since $z(x) = 1 + O(x^{-1})$ as $x \to \infty$, such a b certainly exists.

Thus in every case we have two linearly independent solutions of (1) in the interval $x \geq x_0$, and, if $a < x_0$, these solutions can be extended to the interval $x \geq a$. It remains to show that the formal solutions obtained in sec. 3.2 are asymptotic expansions of the solutions obtained in this section.

3.4. Asymptotic expansions of the solutions

We first remark that

(1) $\displaystyle\int_x^\infty e^{-t} t^{-\nu}\, dt \sim e^{-x} \sum_{m=0}^{\infty} (-1)^m (\nu)_m\, x^{-\nu-m}$ as $x \to \infty$

(2) $\displaystyle\int_b^x e^{t} t^{-\nu}\, dt \sim e^{x} \sum_{m=0}^{\infty} (\nu)_m\, x^{-\nu-m}$ as $x \to \infty$,

where

(3) $(\nu)_0 = 1, \quad (\nu)_r = \nu(\nu+1) \cdots (\nu+r-1),$ $\qquad r = 1, 2, \ldots .$

Both results can be proved by successive integrations by parts, with $g = t^{-\nu}$, $h_{-m} = (\mp 1)^m \exp(\mp t)$ in 2.1(5). In particular,

(4) $\displaystyle\int_x^\infty e^{-t} t^{-\nu}\, dt = O(e^{-x} x^{-\nu})$ as $x \to \infty$

(5) $\displaystyle\int_b^x e^{t} t^{-\nu}\, dt = O(e^{x} x^{-\nu})$ as $x \to \infty$.

Next we prove by induction that *the functions defined by* 3.3(12) *possess asymptotic power series expansions of the form*

(6) $$z_n(x) \sim \sum_{k=n}^{\infty} c_{kn} x^{-k} \qquad \text{as } x \to \infty.$$

This is certainly true for $n = 0$. If it is true for any n, then

$$F(t) z_n(t) \sim \sum_{k=n}^{\infty} a_k t^{-k} \qquad \text{as } t \to \infty.$$

Also

$$z'_{n+1}(x) = \int_x^{\infty} e^{2\omega(t-x)} \left(\frac{x}{t}\right)^{2\rho} F(t) z_n(t)\, t^{-2}\, dt.$$

If ω and ρ are real, the last theorem in sec. 1.4 justifies substitution of the asymptotic expansion of $F(t) z_n(t)$ in the integral, so that

(7) $$z'_{n+1}(x) \sim \sum_{k=n}^{\infty} a_k \int_x^{\infty} e^{2\omega(t-x)} \left(\frac{x}{t}\right)^{2\rho} t^{-k-2}\, dt \qquad \text{as } x \to \infty.$$

If ω or ρ is complex, (7) can be proved by taking the asymptotic expansion of $F(t) z_n(t)$ to a finite number of terms, with a remainder, substituting, and estimating the remainder in (7) by (4). In any event, each integral in (7) possesses an asymptotic power series expansion which can be obtained from (1), and starts with x^{-k-2}. By the third theorem in sec. 1.4, it is permissible to substitute this expansion in (7): $\mu(n) = n$, the uniformity of the asymptotic expansion is trivial, and the series 1.4(5) terminate so that the question of convergence does not arise. Thus we obtain by rearrangement

(8) $$z'_{n+1}(x) \sim \sum_{k=n}^{\infty} b_k x^{-k-2};$$

and by integration of (8) we obtain the asymptotic expansion of z_{n+1}.

By the second theorem of sec. 1.4, it is permissible to substitute (6) in 3.3(13): since $z_n = O(x^{-n})$, the question of uniformity is trivial, and the series 1.4(2) terminate so that the question of their convergence does not arise. Thus we see that $z(x)$ *possesses an asymptotic power series expansion of the form*

(9) $$z(x) \sim \sum_{n=0}^{\infty} c_n x^{-n} \qquad \text{as } x \to \infty.$$

It remains to prove that *the coefficients* c_n *occurring here satisfy* 3.2(7). It follows from 3.3(17) and the corresponding relation for z'', that z' and z'' also possess asymptotic power series expansions. By a result in sec. 1.6 it follows that (9) may be differentiated twice. The resulting asymptotic series must satisfy 3.3(5) formally, and this leads to 3.2(7) for the coefficients. Also, $c_0 = 1$.

We have thus proved that 3.3(18) is represented asymptotically by one of the formal solutions, and we conclude by showing that 3.3(19) is represented by the other. To do this we put

$$(10) \quad y_2(x) = e^{-\omega x} x^{\rho} z_2(x),$$

and have from 3.3(18) and (19)

$$(11) \quad z_2(x) = z(x) \int_b^x e^{2\omega(x-t)} \left(\frac{x}{t}\right)^{-2\rho} [z(t)]^{-2}\, dt.$$

By the choice of b, $z(t)$ is bounded away from zero, and we have also seen that $z(t)$ possesses an asymptotic expansion, (9) with $c_0 = 1$, as $t \to \infty$. Then

$$[z(t)]^{-2} = \sum_{n=0}^{N-1} a_n t^{-n} + O(t^{-N}) \qquad \text{for } t \geq b$$

for some a_n, and

$$\frac{z_2(x)}{z(x)} = \sum_{n=0}^{N-1} a_n \int_b^x e^{2\omega(x-t)} \left(\frac{x}{t}\right)^{-2\rho} t^{-n}\, dt$$

$$+ \int_b^x e^{2\omega(x-t)} \left(\frac{x}{t}\right)^{-2\rho} O(t^{-N})\, dt.$$

The integral under the summation sign can be expanded asymptotically by (2), and the last integral is $O(x^{-N})$ by (5). Since N is an arbitrary positive integer, $z_2(x)/z(x)$ possesses an asymptotic power series. Hence $z_2(x)$ *possesses an asymptotic power series expansion*

$$(12) \quad z_2(x) \sim \sum_{n=0}^{\infty} C_n x^{-n} \qquad \text{as } x \to \infty .$$

By a similar consideration as in the case of z_1, it can be proved that *the coefficients occurring here satisfy a recurrence relation* which differs from 3.2(7) only in that ω, ρ are replaced by $-\omega$, $-\rho$. Thus, 3.3(19) is represented asymptotically by one of the formal solutions.

If ω and ρ are both imaginary, the two fundamental solutions, both of the form 3.3(18), are defined uniquely up to a constant factor: both are bounded and neither approaches zero as $x \to \infty$. In all other cases, one of the fundamental solutions, 3.3(18), approaches zero as $x \to \infty$, and is defined uniquely up to a constant factor: the other, 3.3(19), is unbounded as $x \to \infty$, and is not unique (since it depends on b). In fact,

$$\gamma_1 y_1(x) + \gamma_2 y_2(x) \sim \gamma_2 y_2(x) \qquad \text{as} \qquad x \to \infty \qquad \gamma_2 \neq 0.$$

3.5. Complex variable. Stokes' phenomenon

The results of the preceding sections may be extended to the case of a complex variable x ranging over a sectorial region S,

$$(1) \quad |x| \geq a, \qquad \alpha \leq \arg x \leq \beta.$$

It will be assumed that $q(x)$ is analytic in S, and that 3.3(2) holds, uniformly in $\arg x$, as $x \to \infty$ in S. We then have two formal solutions 3.3(3), where ω satisfies 3.2(5).

The line $\mathrm{Re}[x(-q_0)^{1/2}] = 0$ is called the *critical line* or the *Stokes line*. If $x \to \infty$ along one of the rays of the critical line, the exponential factors in both formal solutions remain bounded, and bounded away from zero. If $x \to \infty$ along any other ray, the leading term in one of the formal solutions increases exponentially.

First we assume that the critical line does not intersect S. Clearly, in this case $\beta - \alpha < \pi$, and we may take ω as that solution of 3.2(5) for which $\mathrm{Re}\ \omega x < 0$ for all x in S. If x varies along any ray $\arg x = \text{const.}$ in S, the results of sections 3.3 and 3.4 hold, and these results can be extended to the sector S as follows. In the integral equation 3.3(10) we always integrate along a ray, so that $\arg x = \arg t$. The boundedness of the kernel then follows for each x in S, and uniformly in x when $\alpha \leq \arg x \leq \beta$. The integral equation can be solved as before, each $z_n(x)$ can be shown to be analytic in S, and $z(x)$ is also analytic in S, since it is the uniform limit of analytic functions. In 3.3(8), b is chosen so that $y_1(x) \neq 0$ when $|x| \geq b$, $\alpha \leq \arg x \leq \beta$. The result is the existence of two solutions, y_1 and y_2, in S which are represented asymptotically by multiples of the formal solutions 3.3 (3). The asymptotic expansions hold uniformly in $\arg x$ as $x \to \infty$, $\alpha \leq \arg x \leq \beta$. Any solution of the differential equation is a linear combination of y_1 and y_2; and its asymptotic expansion follows from the asymptotic expansions of y_1 and y_2.

Next we assume that the critical line intersects S and decomposes it into a finite number of sectors, S_k, $k = 1, \ldots, K$, and certain rays of the line itself. In each of the sectors S_k we have a value ω_k of ω such that $\operatorname{Re} \omega_k x < 0$ for all x in S_k, and in each S_k we have a fundamental system y_{1k}, y_{2k} which is asymptotically represented by the formal solutions. For a fuller discussion of these fundamental systems the reader is referred to Hoheisel (1924). It turns out that the fundamental system belonging to a ray of the critical line may be taken also as a fundamental system for the two sectors separated by that ray. Each of the two solutions is *dominant* (exponentially increasing) in one of the two sectors, and *recessive* (exponentially decreasing) in the other.

Let us consider a solution, $y(x)$, of the differential equation in S. In each of the sectors S_k, y is a linear combination of the two fundamental solutions for that sector; in each of the sectors y will be represented asymptotically by a linear combination of the two formal solutions; but the coefficients may vary from sector to sector. This circumstance was discovered by Stokes, and it is called *Stokes' phenomenon*. The sectors S_k are sometimes called *Stokes sectors*, and the critical rays, *Stokes rays*.

For the determination of the coefficients involved in the expression of $y(x)$ as a linear combination of the formal solutions see Turrittin (1950).

3.6. Bessel functions of order zero

We shall illustrate the results of the last few sections by a brief discussion of the differential equation

$$(1) \quad z'' + x^{-1} z' + z = 0$$

satisfied by Bessel functions of order zero. The change of variable

$$(2) \quad z = x^{-1/2} y,$$

transforms (1) to the standard form

$$(3)\quad y'' + \left(1 + \frac{1}{4x^2}\right) y = 0.$$

This equation is of the form 3.3(1), and in 3.5(1) we may take $a = 0$ and α, β arbitrary.

We obtain formal solutions as in sec. 3.2; equations (5), (6), (7) of that section become

$$\omega^2 + 1 = 0, \qquad \rho = 0, \qquad 2\omega n c_n = (n - 1/2)^2 c_{n-1};$$

and with the abbreviation

$$(4)\quad a_n = \prod_{\nu=1}^{n} \frac{(\nu - 1/2)^2}{2} = \frac{[\Gamma(n + 1/2)]^2}{2^n n!\, \pi}$$

and appropriate choices of c_0, we obtain two formal solutions

$$(5)\quad S_1(x) = (2/\pi)^{1/2} e^{ix - i\pi/4} \sum_{n=0}^{\infty} a_n (-i)^n x^{-n-1/2}$$

$$(6)\quad S_2(x) = (2/\pi)^{1/2} e^{-ix + i\pi/4} \sum_{n=0}^{\infty} a_n i^n x^{-n-1/2}$$

of (1). Between the formal series (5) and (6) the identities

$$(7)\quad S_1(xe^{\pi i}) = -S_2(x), \qquad S_1(xe^{-\pi i}) = S_2(x)$$

$$S_2(xe^{\pi i}) = S_1(x), \qquad S_2(xe^{-\pi i}) = -S_1(x)$$

hold.

Since $\omega = \pm i$, the critical line is the real x axis, and according to the theory outlined in the preceding section, every solution of (1) is represented asymptotically by a linear combination of S_1 and S_2 in any sector which is entirely in the upper (or lower) half-plane. As the real axis is crossed, the coefficients may change. We shall see that such changes actually occur.

It can be verified by substitution that (1) is satisfied by the *Bessel function of the first kind of order zero*

$$(8)\quad J_0(x) = \sum_{m=0}^{\infty} \frac{(-1)^m}{(m!)^2} (\tfrac{1}{2}x)^{2m}$$

which is an even entire function of x. Poisson's integral representation

$$(9)\quad J_0(x) = \frac{1}{\pi}\int_{-1}^{1} e^{ixu} (1-u^2)^{-\frac{1}{2}}\, du$$

may be verified by expanding the exponential function and integrating term-by-term.

Let us assume for the moment that Re $x > 0$. Then e^{ixu} vanishes exponentially as Im $u \to +\infty$, and we may break up the integral in (9) according to

$$\int_{-1}^{1} = -\int_{1}^{1+i\infty} + \int_{-1}^{-1+i\infty}$$

In the first integral we put

$$u = 1 + it, \qquad 1 - u = te^{-i\pi/2}, \qquad 1 + u = 2 + it,$$

and in the second,

$$u = -1 + it, \qquad 1 - u = 2 - it, \qquad 1 + u = te^{i\pi/2},$$

thus obtaining two functions which are constant multiples of

$$(10)\quad H_0^{(1)}(x) = \frac{2}{\pi} e^{ix - i\pi/4} \int_0^{\infty} e^{-xt} t^{-\frac{1}{2}} (2+it)^{-\frac{1}{2}}\, dt$$

$$H_0^{(2)}(x) = \frac{2}{\pi} e^{-ix + i\pi/4} \int_0^{\infty} e^{-xt} t^{-\frac{1}{2}} (2-it)^{-\frac{1}{2}}\, dt.$$

The functions defined by (10) are known as *Bessel functions of the third kind*, or *Hankel functions*, *of order zero*. These functions are defined by (10) for Re $x > 0$, but their domains of definition can be extended to $-\pi < \arg x < 2\pi$ in the case of $H_0^{(1)}$, and to $-2\pi < \arg x < \pi$ in the case of $H_0^{(2)}$, by rotating the path of integration as in sec. 2.2.

It can be shown that $H_0^{(1)}$ and $H_0^{(2)}$ are also solutions of (1). Clearly

$$(11)\quad J_0(x) = \frac{1}{2} H_0^{(1)}(x) + \frac{1}{2} H_0^{(2)}(x),$$

and a closer investigation reveals that both $H_0^{(1)}$ and $H_0^{(2)}$ have logarithmic singularities at the origin. The knowledge of these singularities leads to a definition of the Hankel functions for all values of $\arg x$.

The integrals representing Hankel functions are Laplace integrals, and their asymptotic expansions for large x may be obtained by means of the last theorem in sec. 2.2, the result being

(12) $H_0^{(1)}(x) \sim S_1(x)$, uniformly in $\arg x$, as $x \to \infty$,
$-\pi + \epsilon \le \arg x < 2\pi - \epsilon, \quad \epsilon > 0$

(13) $H_0^{(2)}(x) \sim S_2(x)$, uniformly in $\arg x$, as $x \to \infty$,
$-2\pi + \epsilon \le \arg x \le \pi - \epsilon, \quad \epsilon > 0.$

From (11) we then have

(14) $2J_0(x) \sim S_1(x) + S_2(x)$, uniformly in $\arg x$, as $x \to \infty$,
$-\pi + \epsilon \le \arg x \le \pi - \epsilon, \quad \epsilon > 0.$

In the last equation we have used the symbol $\sim$ in a somewhat unusual manner, in that the right-hand side is not one asymptotic expansion but the sum of two. A justificiation of this use is based on the circumstance that on the real axis the two expansions are of the same order and may be combined into a single expansion (which then is not an asymptotic power series), while in the upper [lower] half-plane $S_1(x)$ [$S_2(x)$] is recessive and may be omitted.

We have thus obtained the asymptotic expansion of $J_0(x)$ in the whole plane with the exception of a narrow sector around the negative real axis. To obtain asymptotic expansions valid in sectors including the negative real axis, we remark that it follows from (8) and (14) that

$$2J_0(x) = 2J_0(xe^{\pi i}) \sim S_1(xe^{\pi i}) + S_2(xe^{\pi i})$$

as $x \to \infty$ and $-\pi + \epsilon \le \arg(xe^{\pi i}) \le \pi - \epsilon$, so that

(15) $2J_0(x) \sim S_1(x) - S_2(x)$, uniformly in $\arg x$, as $x \to \infty$,
$-2\pi + \epsilon \le \arg x \le -\epsilon$

by (7); and similarly $J_0(x) = J_0(xe^{-\pi i})$ and

(16) $2J_0(x) \sim -S_1(x) + S_2(x)$, uniformly in $\arg x$, as $x \to \infty$,
$\epsilon \le \arg x \le 2\pi - \epsilon.$

A comparison of (14), (15), and (16) shows the Stokes' phenomenon. The rays excluded by narrow sectors are the Stokes rays. At first it may seem strange that the sectors of validity of these asymptotic expansions overlap, but there is no contradiction involved in this. The regions of validity of (14) and (15) have the common part $-\pi + \epsilon \leq \arg x \leq -\epsilon$; in this common part $S_2(x)$ is recessive so that the right-hand sides of (14) and (15) are asymptotically equal. Thus, the coefficients of the formal series jump in sectors where these series are dominated by the other series.

REFERENCES

Hoheisel, G.K.H., 1924: *J. Reine Angew. Math.* 153, 228-248.

Horn, Jakob, 1915: *Jber. Deutsch. Math. Verein.* 24, 309-329.

Horn, Jakob, 1916: *Jber. Deutsch. Math. Verein.* 25, 74-83, 301-325.

Horn, Jakob, 1919: *Math. Z.* 3, 265-313.

Horn, Jakob, 1920: *Math. Z.* 8, 100-114.

Horn, Jakob, 1924: *Math. Z.* 21, 85-95.

Ince, E.L., 1927: *Ordinary differential equations.* Longmans, Green and Co.

Kamke, E.W.H., 1944: *Differentialgleichungen. Lösungsmethoden und Lösungen.* Third Edition, Leipzig.

Poincaré, Henri, 1886: *Acta Math.* 8, 295-344.

Poole, E.G.C., 1936: *Introduction to the theory of linear differential equations.* Oxford.

Sternberg, Wolfgang, 1920: *Math. Ann.* 81, 119-186.

Tricomi, F.G., 1953: *Equazioni differenziali.* Second edition, Torino.

Trjitzinsky, W.J., 1933: *Acta Math.* 62, 167-226.

Turrittin, H.L., 1950: *Trans. Amer. Math. Soc.* 68, 304-329.

Wasow, W.R., 1953: *Introduction to the asymptotic theory of ordinary linear differential equations.* Working paper, National Bureau of Standards.

CHAPTER IV

DIFFERENTIAL EQUATIONS WITH A LARGE PARAMETER

In this chapter we describe briefly the asymptotic theory of ordinary homogeneous linear differential equations of the second order containing a large parameter. For this theory and its extensions to equations of higher orders, and to systems of differential equations of the first order, see Ince (1927, p. 270 ff.), Kamke (1944, p. 62 ff., 102 ff., 137 ff., 213 ff.), Wasow (1953), the references given in these works, and the references given at the end of this chapter.

As in sec. 3.2, we may transform the differential equation to standard form

$$y'' + q(x, \lambda)y = 0,$$

where x is a real or complex variable, and λ is a real or complex parameter. We shall investigate the behavior of the solutions of this differential equation as $\lambda \to \lambda_0$, and without loss of generality we take $\lambda_0 = \infty$.

The reader will be expected to be familiar with the basic theorems regarding the dependence of solutions of a differential equation on parameters occurring in the equation (see, for instance, Kamke, 1930, sec. 17, sec. 8).

4.1. Liouville's problem

In the course of his classical investigations of the Sturm-Liouville problem, Liouville discussed the behavior of solutions of the differential equation

$$(1) \quad y'' + [\lambda^2 p(x) + r(x)]y = 0$$

as $\lambda \to \infty$. Here x is a real variable, $a \le x \le b$, $p(x)$ is positive and twice continuously differentiable, and $r(x)$ is continuous, for $a \le x \le b$. Liouville's procedure may be summarized as follows.

New variables, ξ and η, are introduced by the substitution

(2) $\xi = \int [p(x)]^{½}\,dx, \qquad \eta = [p(x)]^{¼}\,y,$

which carries the interval $a \le x \le b$ into $\alpha \le \xi \le \beta$, and the differential equation (1) into

(3) $$\frac{d^2\eta}{d\xi^2} + \lambda^2\eta = \rho(\xi)\,\eta,$$

where

(4) $$\rho(\xi) = \frac{1}{4}\frac{p''}{p^2} - \frac{5}{16}\frac{p'^2}{p^3} - \frac{r}{p}$$

is a continuous function of ξ, $\alpha \le \xi \le \beta$. ($p' = dp/dx$, etc.)

By a method similar to that employed in sec. 3.3 it can now be shown that solutions of (3) satisfy the Volterra integral equation

(5) $$\eta(\xi) = c_1 \cos\lambda\xi + c_2 \sin\lambda\xi + \lambda^{-1}\int_\gamma^\xi \sin\lambda(\xi - t)\,\rho(t)\,\eta(t)\,dt,$$

where $\alpha \le \gamma \le \beta$ and c_1, c_2 are arbitrary. $\eta(\xi)$ and $c_1 \cos\lambda\xi + c_2 \sin\lambda\xi$ have the same value, and the same derivative, at $\xi = \gamma$.

The solution of (5) can be obtained by successive approximations in the form

(6) $$\eta(\xi, \lambda) = \sum_{n=0}^{\infty} \eta_n(\xi, \lambda),$$

where

$$\eta_0(\xi, \lambda) = c_1 \cos\lambda\xi + c_2 \sin\lambda\xi$$

$$\eta_{n+1}(\xi, \lambda) = \lambda^{-1}\int_\gamma^\xi \sin\lambda(\xi - t)\,\rho(t)\,\eta_n(t, \lambda)\,dt \qquad n = 0, 1, \ldots .$$

If $|\rho(\xi)| \le A$, it is easy to prove by induction that

$$|\eta_n(\xi, \lambda)| \le \frac{|c_1| + |c_2|}{n!}\,\frac{A^n\,|\xi - \gamma|^n}{\lambda^n} \qquad n = 1, 2, \ldots ,$$

and in the case of a finite interval (α, β) it follows that (6) is uniformly convergent for $\alpha \leq \xi \leq \beta$, $\lambda \geq \lambda_1 > 0$, and is also an asymptotic expansion of $\eta(\xi, \lambda)$ as $\lambda \to \infty$.

The η_n are very difficult to compute. Other approximations for large λ may be obtained from formal solutions (which are divergent in general). There are two methods. One uses formal expansions

$$\sum_{n=0}^{\infty} \alpha_n(x)\,\lambda^{-n} \cos\lambda\xi + \sum_{n=0}^{\infty} \beta_n(x)\,\lambda^{-n} \sin\lambda\xi$$

for $y(x, \lambda)$, while the other uses formal expansions

$$i\lambda\xi + \sum_{n=0}^{\infty} \gamma_n(x)\,\lambda^{-n}$$

for $\log y(x, \lambda)$. In the second method, y is a solution without zeros. In either method an approximation to y is constructed by taking a partial sum of the formal expansion; and this approximation is compared with y by means of an integral equation. Either of these two methods reduces to Liouville's process if the partial sum in question consists of one term.

4.2. Formal solutions

Instead of 4.1 (1), we shall discuss the slightly more general differential equation

(1) $\quad y'' + q(x, \lambda)y = 0.$

If $q(x, \lambda)$ is a formal power series in λ^{-1} with coefficients which depend on x, then two linearly independent solutions of (1) may also be represented by formal power series in λ^{-1}. On the other hand, if the formal expansion of q in powers of λ contains positive powers of λ, then the formal expansion of y will be a Laurent series. Nevertheless, we shall see that in the case that $q(x, \lambda)$, as a function of λ, has a pole at $\lambda = \infty$, we can construct formal solutions which are analogous to the normal and subnormal solutions of sec. 3.2.

In (1), we shall assume that $q(x, \lambda)$ is of the form

(2) $\quad \sum_{n=0}^{\infty} q_n(x)\,\lambda^{2k-n},$

where the $q_n(x)$ are independent of λ, and k is a positive integer. We shall also assume that $q_0(x)$ does not vanish in the interval (or simply-connected region in the complex plane) over which x varies.

Corresponding to the two methods mentioned at the end of sec. 4.1, we shall obtain two kinds of formal solutions of (1). The first one of these is of the form

$$(3)\quad \sum_{n=0}^{\infty} a_n(x)\,\lambda^{-n} \exp\Big[\sum_{\nu=0}^{k-1} \beta_\nu(x)\,\lambda^{k-\nu}\Big].$$

In substituting (3) in (1), we use the convention $q_n = 0$, $a_n = 0$ for $n = -1, -2, -3, \ldots$, and $\beta_\nu = 0$ for $\nu = -1, -2, \ldots$ and also for $\nu = k$, $k+1, \ldots$.

All summations may then be extended over all integers, and we obtain

$$[\Sigma\, \beta_\nu''\, \lambda^{k-\nu} + (\Sigma\, \beta_\nu'\, \lambda^{k-\nu})^2]\, \Sigma\, a_n\, \lambda^{-n} + 2\,\Sigma\, \beta_\nu'\, \lambda^{k-\nu}\, \Sigma\, a_n'\, \lambda^{-n}$$
$$+ \Sigma\, a_n''\, \lambda^{-n} + \Sigma\, q_n\, \lambda^{2k-n}\, \Sigma\, a_n\, \lambda^{-n} = 0.$$

Picking out the coefficient of λ^{2k-n}

$$(4)\quad \sum_m a_{n-m}\Big(q_m + \sum_\nu \beta_\nu'\, \beta_{m-\nu}'\Big) + \sum_m a_{n-m}\, \beta_{m-k}''$$
$$+ 2\sum_m a_{n-m}'\, \beta_{m-k}' + a_{n-2k}'' = 0$$

for all integer values of n.

The first non-vacuous condition arises when $n = 0$. If we set $n = 0, 1, \ldots, k-1$ in (4) we obtain

$$q_m + \sum_\nu \beta_\nu'\, \beta_{m-\nu}' = 0 \qquad m = 0, 1, \ldots, k-1$$

or

$$(5)\quad \beta_0'^{\,2} + q_0 = 0$$

$$(6)\quad 2\beta_0'\, \beta_m' + q_m + \sum_{\nu=1}^{m-1} \beta_\nu'\, \beta_{m-\nu}' = 0 \qquad m = 1, \ldots, k-1.$$

In consequence of these relations, we may restrict summation to $m \geq k$ in the first sum in (4). For $n = k$ in (4) we have

$$(7)\quad 2a_0'\, \beta_0' + a_0\Big(\beta_0'' + q_k + \sum_{\nu=1}^{k-1} \beta_\nu'\, \beta_{k-\nu}'\Big) = 0,$$

and when we replace n by $k + n$ in (4),

$$(8)\quad 2\alpha_n' \beta_0' + \alpha_n(\beta_0'' + q_k + \sum_{\nu=1}^{k-1} \beta_\nu' \beta_{k-\nu}')$$

$$+ \sum_{m=1}^{n} \alpha_{n-m} (\beta_m'' + q_{k+m} + \sum_{\nu=m+1}^{k-1} \beta_\nu' \beta_{k+m-\nu}')$$

$$+ 2 \sum_{m=1}^{n} \alpha_{n-m}' \beta_m' + \alpha_{n-k}'' = 0 \qquad n = 1, 2, \dots .$$

We thus see that (3) satisfies (1) formally, provided that the α_n and the β_ν satisfy (5) to (8). In these equations, empty sums (i.e., sums whose upper limit of summation is less than the lower limit) are interpreted as zero. Since $q_0 \neq 0$, we may choose a branch of $[-q_0(x)]^{1/2}$, and then (5) determines β_0 up to an additive constant. Moreover, $\beta_0' \neq 0$, and hence (6) determines $\beta_1, \dots, \beta_{k-1}$ recurrently, up to an additive constant in each. (7) determines α_0 up to a constant factor, and (8) determines $\alpha_1, \alpha_2, \dots$ recurrently, up to an additive constant multiple of α_0 in each. Corresponding to the two branches of $(-q_0)^{1/2}$, we obtain two formal solutions of the form (3).

A second type of formal solution is

$$(9)\quad \exp[\sum_{n=0}^{\infty} \beta_n(x) \lambda^{k-n}].$$

Substituting (9) in (1) we obtain

$$\Sigma \beta_n'' \lambda^{k-n} + (\Sigma \beta_n' \lambda^{k-n})^2 + \Sigma q_n \lambda^{2k-n} = 0,$$

and comparing coefficients of λ^{2k-n},

$$(10)\quad \beta_0'^2 + q_0 = 0$$

$$(11)\quad 2 \beta_0' \beta_n' + q_n + \sum_{m=1}^{n-1} \beta_m' \beta_{n-m}' = 0 \qquad n = 1, \dots, k-1$$

$$(12)\quad 2 \beta_0' \beta_n' + q_n + \sum_{m=1}^{n-1} \beta_m' \beta_{n-m}' + \beta_{n-k}'' = 0 \qquad n = k, k+1, \dots .$$

There are two linearly independent formal solutions of this type.

The connection between these two types of formal solutions is fairly obvious. Equations (10) and (11) are identical with (5) and (6), and $\Sigma\, a_n \lambda^{-n}$ is the formal expansion of

$$\exp\left(\sum_{n=k}^{\infty} \beta_n \lambda^{k-n}\right).$$

Throughout this discussion we have assumed that $q(x, \lambda)$, as a function of λ, has a pole of even order at $\lambda = \infty$. If the pole is of odd order, then no solution of the form (3) or (9) exists, and instead of powers of λ we must expand in powers of $\lambda^{1/2}$.

4.3. Asymptotic solutions

We shall now show that under certain assumptions, the differential equation 4.2(1) possesses a fundamental system of solutions which are represented asymptotically by the formal solutions obtained in the preceding section. It does not matter whether we compare solutions of 4.2(1) with

$$\sum_{n=0}^{N-1} a_n(x)\, \lambda^{-n} \exp\left[\sum_{\nu=0}^{k-1} \beta_\nu(x)\, \lambda^{k-\nu}\right],$$

where the a_n and β_ν satisfy 4.2(5) to (8), or with

$$\exp\left[\sum_{n=0}^{k+N-1} \beta_n(x)\, \lambda^{k-n}\right],$$

where the β_n satisfy 4.2(10) to (12), for the a's and β's can be so chosen that the ratio of these two expressions is $1 + O(\lambda^{-N})$.

We fix a positive integer N, and set

$$(1) \quad Y_j(x) = \exp\left[\sum_{n=0}^{2k+N-1} \beta_{nj}(x)\, \lambda^{k-n}\right] \qquad j = 1, 2,$$

where $\beta'_{01} = -\beta'_{02}$, and for each j, the β_{nj} satisfy 4.2(10) to (12). These coefficients are completely determined by $q_0, \ldots, q_{2k+N-1}$, and certain derivatives of these functions, and we shall say that the q_n are sufficiently often differentiable if all the derivatives entering the determination of the β_{nj}, $n = 0, \ldots, 2k + N - 1$, exist and are continuous functions of x. We let x vary over a bounded and closed interval I: $a \leq x \leq b$, and λ, over a sectorial domain S: $|\lambda| \geq \lambda_1$, $\phi_0 \leq \arg \lambda \leq \phi_1$. The theorem to be proved is as follows.

If for each fixed λ *in* S, $q(x, \lambda)$ *is a continuous function of* x *over* I; *if*

(2) $$q(x, \lambda) = \sum_{n=0}^{2k+N-1} q_n(x)\, \lambda^{2k-n} + O(\lambda^{-N}),$$

uniformly in x *and* $\arg \lambda$, *as* $\lambda \to \infty$ *in* S, *where the* $q_n(x)$ *are sufficiently often differentiable in* I, *and*

(3) $$\operatorname{Re}\{\lambda^k [-q_0(x)]^{1/2}\} \neq 0$$

when λ *is in* S *and* x *in* I, *then the differential equation*

(4) $$y'' + q(x, \lambda)\, y = 0$$

possesses a fundamental system of solutions, $y_1(x)$ *and* $y_2(x)$, *so that*

(5) $$y_j(x) = Y_j(x)\,[1 + O(\lambda^{-N})],$$
$$y'_j(x) = Y'_j(x)\,[1 + O(\lambda^{-N})],$$

uniformly in x *and* $\arg \lambda$, *as* $\lambda \to \infty$ *in* S.

We shall prove this theorem by a method analogous to that used in sec. 3.3. By (3) and 4.2(10) we may choose β_{01} and β_{02} so that for each λ in S, $\operatorname{Re}[\lambda^k \beta_{01}(x)]$ is an increasing, and $\operatorname{Re}[\lambda^k \beta_{02}(x)]$ a decreasing, function of x. It then follows from (1) that for each sufficiently large λ in S, $|Y_1(x)|$ is an increasing, and $|Y_2(x)|$ a decreasing, function of x.

To establish the existence and the asymptotic property of $y_1(x)$, we substitute

(6) $$y_1(x) = Y_1(x)\, z(x)$$

in (4) and obtain

(7) $$z'' + 2\frac{Y'_1}{Y_1} z' + F(x, \lambda)\, z = 0,$$

where

(8) $$F(x, \lambda) = \frac{Y''_1}{Y_1} + q = \sum_{n=0}^{2k+N-1} \beta''_{n1} \lambda^{k-n} + \Big(\sum_{n=0}^{2k+N-1} \beta'_{n1}\, \lambda^{k-n}\Big)^2 + q = O(\lambda^{-N})$$

uniformly in x and arg λ, as $\lambda \to \infty$ in S, by (2) and 4.2(10) to (12). Equation (7) may be written as

$$\frac{d}{dx}\left[Y_1^2(x)\frac{dz}{dx}\right] + Y_1^2(x)\,F(x,\lambda)\,z = 0,$$

and by two successive integrations, and a suitable choice of the constants of integration, we obtain

$$(9) \quad z(x) = 1 - \int_a^x K(x,t)\,F(t,\lambda)\,z(t)\,dt,$$

where

$$K(x,t) = \int_t^x Y_1^2(t)\,Y_1^{-2}(s)\,ds.$$

Since $|Y_1(x)|$ is an increasing function, we have $|Y_1(t)| \le |Y_1(s)|$, and

$$|K(x,t)| \le (b-a) \qquad a \le t \le x \le b.$$

The existence of $z(x)$ now follows from the general theory of Volterra integral equations, or can be established by successive approximations. From (8) and (9), $z(x) = 1 + O(\lambda^{-N})$, uniformly in x and arg λ, as $\lambda \to \infty$ in S. Moreover, $z(x)$ is differentiable,

$$z'(x) = -\int_a^x Y_1^2(t)\,Y_1^{-2}(x)\,F(t,\lambda)\,z(t)\,dt = O(\lambda^{-N}),$$

and

$$y_1'(x) = Y_1'(x)\left[z(x) + \frac{Y_1(x)}{Y_1'(x)}\,z'(x)\right] = Y_1'(x)\,[1 + O(\lambda^{-N})].$$

This proves (5) for $j = 1$. The proof for $j = 2$ is similar, except that b, rather than a, must be chosen as the fixed limit of integration in the integral equation.

4.4. Application to Bessel functions

We shall now apply the methods of the last two sections to prove the asymptotic formulas*

$$(1) \quad J_\lambda(\lambda \operatorname{sech}\beta) \sim (2\pi\lambda\tanh\beta)^{-1/2}\exp(\lambda\tanh\beta - \lambda\beta) \quad \text{as} \quad \lambda \to \infty$$

$$(2) \quad H_\lambda^{(1,2)}(\lambda\sec\beta) \sim \left(\frac{1}{2}\pi\lambda\tan\beta\right)^{-1/2}\exp[\pm i(\lambda\tan\beta - \lambda\beta - \pi/4)] \quad \text{as } \lambda \to \infty.$$

* The symbol β used in these formulas must not be confused with the function $\beta(x)$ to be used below.

Equation (1) holds for $\beta > 0$, uniformly in β if $0 < \beta_1 \le \beta \le \beta_2 < \infty$. Equation (2) holds for $0 < \beta < \pi/2$, uniformly in β if $0 < \epsilon \le \beta \le \pi/2 - \epsilon$; and in this equation (and similarly later in (18)) the upper sign holds for $H_\lambda^{(1)}$, the lower sign for $H_\lambda^{(2)}$. Both results may be derived from integral representations of Bessel functions by means of the method of steepest descents (Watson, 1922, sec. 8.4, 8.41).

The functions

$$x^{\frac{1}{2}} J_\lambda(\lambda x), \quad x^{\frac{1}{2}} Y_\lambda(\lambda x), \quad x^{\frac{1}{2}} H_\lambda^{(1)}(\lambda x), \quad x^{\frac{1}{2}} H_\lambda^{(2)}(\lambda x)$$

are solutions of the differential equation

$$(3) \quad y'' + [\lambda^2 - (\lambda^2 - \tfrac{1}{4}) x^{-2}] y = 0.$$

This equation is of the form 4.3 (4) with

$$k = 1, \quad q_0(x) = 1 - x^{-2}, \quad q_2(x) = (2x)^{-2},$$

all the other $q_n(x)$ vanishing identically. The points $x = 0, \infty$ are singular points of (3), and $x = 1$ is a so-called *transition point* at which the condition 4.3 (3) is violated for any value of λ. On any interval $a \le x \le b$ which does not include any singularity or transition point, the theorem of sec. 4.3 will yield the general form of the asymptotic solutions but it will fail to indicate the expression of y_1 and y_2 in terms of the standard Bessel functions. In order to identify our solutions in terms of Bessel functions, we shall extend the interval to one of the singular points of (3). Since this case is not covered by the theorem of sec. 4.3, we shall use the methods rather than the results of the preceding sections.

Let us first discuss (3) on the interval $0 \le x \le b < 1$. From 4.2 (5) we have

$$\beta_0'(x) = \pm (x^{-2} - 1)^{\frac{1}{2}},$$

and hence

$$(4) \quad \pm \beta_0(x) = \beta(x) = \int (x^{-2} - 1)^{\frac{1}{2}} \, dx = (1 - x^2)^{\frac{1}{2}} + \log \frac{x}{1 + (1 - x^2)^{\frac{1}{2}}}.$$

From 4.2 (7),

$$2a_0' \beta_0' + a_0 \beta_0'' = 0,$$

so that

(5) $a_0(x) = [\beta'(x)]^{-\frac{1}{2}} = a(x) = x^{\frac{1}{2}}(1-x^2)^{-\frac{1}{4}}$.

With a and β so defined we form the functions

(6) $Y_1(x) = a(x)\, e^{\lambda\beta(x)}, \quad Y_2(x) = a(x)\, e^{-\lambda\beta(x)}$,

which correspond to the leading terms of the formal solutions.

The integral equation for $z = y_1/Y_1$ is

(7) $z(x) = 1 - \int_0^x K(x,t)\, F(t,\lambda)\, z(t)\, dt.$

By a straightforward computation we find

$$(8) \quad F(x,\lambda) = \frac{Y_1''}{Y_1} + \lambda^2\left(1 - \frac{1}{x^2}\right) + \frac{1}{4x^2} = \frac{4+x^2}{4(1-x^2)^2},$$

so that F is bounded on the interval. Moreover, $a^{-2} = \beta'$ from (5), and

$$(9) \quad K(x,t) = \int_t^x \left[\frac{a(t)}{a(s)}\right]^2 \exp\{2\lambda[\beta(t) - \beta(s)]\}\, ds$$

$$= \frac{a^2(t)}{2\lambda}\,(1 - \exp\{2\lambda[\beta(t) - \beta(x)]\}).$$

Now, $\beta(x)$ is an increasing function of x, $\beta(t) - \beta(x) \le 0$ for $0 < t \le x \le b < 1$, and the exponential function will be bounded if Re $\lambda \ge 0$. Also $a^2(t)$ is bounded. We thus arrive at the estimate

$$(10) \quad |F(t,\lambda)\, K(x,t)| \le \frac{C}{|\lambda|}$$

for Re $\lambda \ge 0$, $0 < t \le x \le b < 1$. Here C is independent of λ, x, t.

We are now ready to solve the integral equation (7) by successive approximations in the form

$$z_0(x) = 1$$

$$z_{n+1}(x) = \int_0^x K(x,t)\, F(t,\lambda)\, z_n(t)\, dt \qquad n = 0, 1, \dots$$

$$z(x) = \sum_{n=0}^{\infty} z_n(x).$$

From (10) it is easy to prove by induction that

$$|z_n(x)| \leq \frac{1}{n!}\left(\frac{Cx}{|\lambda|}\right)^n,$$

so that the series defining z converges uniformly in x and λ if λ is bounded away from zero; $z(x)$ satisfies the integral equation, is twice continuously differentiable, satisfies the appropriate differential equation, and

$$z(x) = 1 + O\left(\frac{x}{\lambda}\right).$$

This establishes the existence of the solution y_1 of (3) which has the property

$$\text{(11)} \quad y_1(x) = Y_1(x)\left[1 + O\left(\frac{x}{\lambda}\right)\right], \qquad 0 < x \leq b < 1, \quad \operatorname{Re}\lambda \geq 0.$$

It remains to show that this solution is a multiple of $J_\lambda(\lambda x)$.

Since y_1 is a solution of (3),

$$\text{(12)} \quad x^{-1/2} y_1(x) = c_1(\lambda)\, J_\lambda(\lambda x) + c_2(\lambda)\, Y_\lambda(\lambda x).$$

Now, fix λ, and let $x \to 0$. It is well known (Watson, 1922, p. 40, p. 64) that

$$J_\lambda(\lambda x) \sim \frac{(\lambda x/2)^\lambda}{\Gamma(\lambda+1)} \qquad \text{as} \quad x \to 0$$

$$Y_\lambda(\lambda x) \sim \frac{(\lambda x/2)^\lambda}{\Gamma(\lambda+1)} \cot \lambda\pi - \frac{(\lambda x/2)^{-\lambda}}{\Gamma(1-\lambda)} \operatorname{cosec} \lambda\pi \qquad \text{as} \quad x \to 0,$$

and it follows from (11), (6), (5), and (4) that *

$$x^{-1/2} y_1(x) \sim x^{-1/2} Y_1(x) \sim e^{\lambda\beta(x)} \sim (x/2)^\lambda e^\lambda \qquad \text{as } x \to 0.$$

Making $x \to 0$ in (12), we see that

$$c_1(\lambda) = e^\lambda \lambda^{-\lambda}\Gamma(\lambda+1), \qquad c_2(\lambda) = 0,$$

* Note that Y_1 is the approximate solution, Y_λ a Bessel function of the second kind.

and by Stirling's formula,

$$c_1(\lambda) = (2\pi\lambda)^{1/2}\left[1 + O\left(\frac{1}{\lambda}\right)\right]$$

We thus find that

$$(13)\quad J_\lambda(\lambda x) = \frac{\lambda^\lambda e^{-\lambda} x^{-1/2}}{\Gamma(\lambda+1)}\, Y_1(x)\left[1 + O\left(\frac{x}{\lambda}\right)\right]$$

$$= (2\pi\lambda x)^{-1/2}\, Y_1(x)\left[1 + O\left(\frac{1}{\lambda}\right)\right]$$

when $0 < x \leq b < 1$ and $\operatorname{Re}\lambda \geq 0$. If we put $x = \operatorname{sech}\beta$ and take λ positive in the latter form, we obtain (1).

Let us now turn to the discussion of (3) on the interval $1 < a \leq x < \infty$. In this case

$$(14)\quad a(x) = x^{1/2}(x^2 - 1)^{-1/4} = [-i\beta'(x)]^{-1/2}$$

$$\beta(x) = i\int(1 - x^{-2})^{1/2}\,dx = i(x^2 - 1)^{1/2} - i\cos^{-1} x^{-1},$$

where $\cos^{-1}$ denotes the principal value of the inverse cosine, in particular $\cos^{-1} x^{-1} \to \pi/2$ as $x \to \infty$. The comparison functions are again of the form (6), with a and β defined in (14). The integral equation for $z = y_1/Y_1$ is

$$(15)\quad z(x) = 1 + \int_x^\infty K(x, t)\, F(t, \lambda)\, z(t)\, dt.$$

Equation (8) holds, and shows that $F(t, \lambda) = O(t^{-2})$ for $t \geq b > 1$ and all λ. Since $a^{-2} = -\beta'$ in this case, the evaluation of the integral in (9) leads to

$$K(x, t) = -\frac{a^2(t)}{2\lambda}\left(1 - \exp\{2\lambda[\beta(t) - \beta(x)]\}\right).$$

Now, $-i\beta$ is an increasing function of x, and $t \geq x$, so that the exponential function will be bounded if $\operatorname{Im}\lambda \geq 0$. Also $a^2(t)$ is bounded, and we have the estimate

(16) $$|F(t, \lambda)\, K(x, t)| \leq \frac{C}{|\lambda|\, t^2}$$

for $\operatorname{Im} \lambda \geq 0$, $1 < a \leq x \leq t$, C being independent of λ, x, t.

We now set

$$z_0(x) = 1$$

$$z_{n+1}(x) = \int_x^\infty K(x, t)\, F(t, \lambda)\, z_n(t)\, dt \qquad n = 0, 1, \ldots$$

$$z(x) = \sum_{n=0}^{\infty} z_n(x).$$

From (16) it is easy to prove by induction that

$$|z_n(x)| \leq \frac{1}{n!} \left(\frac{C}{|\lambda|\, x} \right)^n,$$

so that the series defining z converges uniformly if λ is bounded away from zero; $z(x)$ satisfies the integral equation, is twice continuously differentiable, satisfies the appropriate differential equation, and

$$z(x) = 1 + O\left(\frac{1}{\lambda x} \right).$$

This establishes the existence of a solution y_1 of (3) which has the property

(17) $$y_1(x) = Y_1(x) \left[1 + O\left(\frac{1}{\lambda x} \right) \right] \qquad 1 < a \leq x < \infty, \quad \operatorname{Im} \lambda \geq 0.$$

It remains to show that this solution is a multiple of $H_\lambda^{(1)}(\lambda x)$.

Since y_1 is a solution of (3),

(18) $$x^{-1/2}\, y_1(x) = c_1(\lambda)\, H_\lambda^{(1)}(\lambda x) + c_2(\lambda)\, H_\lambda^{(2)}(\lambda x).$$

Let us fix λ and make $x \to \infty$. It is well known (Watson, 1922, sec. 7.2) that for $0 \leq \arg \lambda < \pi$

$$H_\lambda^{(1,2)}(\lambda x) \sim \left(\frac{2}{\pi \lambda x} \right)^{1/2} \exp\left[\pm i \left(\lambda x - \frac{\lambda \pi}{2} - \frac{\pi}{4} \right) \right] \text{ as } \qquad x \to \infty,$$

and it follows from (17) and (14) that

$$y_1(x) \sim Y_1(x) \sim \exp[\lambda\beta(x)] \sim \exp[i\lambda x - i\lambda\pi/2] \quad \text{as} \quad x \to \infty.$$

Making $x \to \infty$ in (18) we see that

$$c_1(\lambda) = (1/2\pi\lambda)^{1/2}\, e^{i\pi/4}, \qquad c_2(\lambda) = 0.$$

If $\arg \lambda = \pi$, we must use a slightly different asymptotic formula for $H_\lambda^{(2)}$, but the conclusion remains the same. We thus have

$$(19)\quad H_\lambda^{(1)}(\lambda x) = (1/2\pi\lambda x)^{-1/2}\, e^{-i\pi/4}\, Y_1(x)\left[1 + O\left(\frac{1}{\lambda x}\right)\right]$$

where $1 < a \leq x < \infty$ and $\operatorname{Im} \lambda \geq 0$. By a similar proof

$$(20)\quad H_\lambda^{(2)}(\lambda x) = (1/2\pi\lambda x)^{-1/2}\, e^{i\pi/4}\, Y_2(x)\left[1 + O\left(\frac{1}{\lambda x}\right)\right]$$

when $1 < a \leq x < \infty$ and $\operatorname{Im} \lambda \leq 0$. If we take λ positive and put $x = \sec \beta$, we obtain (2).

4.5. Transition points

Let us consider again Liouville's differential equation

$$(1)\quad y'' + [\lambda^2 p(x) + r(x)]\, y = 0$$

with large positive λ. As in sec. 4.1, x is a real variable, $a \leq x \leq b$, $p(x)$ is real and twice continuously differentiable, and $r(x)$ is continuous, for $a \leq x \leq b$. Instead of assuming $p(x)$ to be positive, we now assume that $p(x)$ has a zero in (a, b). A zero of $p(x)$ will be called a *transition point* of the differential equation (1). For the sake of definiteness, we assume that $p(x)$ has a simple zero at $x = c$, and no other zero in $a \leq x \leq b$, and also that $p'(c) > 0$, so that $p(x) < 0$ when $a \leq x < c$, and $p(x) > 0$ when $c < x \leq b$.

We have seen that in the interval $c + \epsilon \leq x \leq b$, $\epsilon > 0$, where $p(x)$ is positive, solutions of (1) are asymptotically of the form

$$(2)\quad c_1 [p(x)]^{-1/4} \cos\{\lambda \int [p(x)]^{1/2}\, dx\} + c_2 [p(x)]^{-1/4} \sin\{\lambda \int [p(x)]^{1/2}\, dx\},$$

and it can be shown similarly that in the interval $a \leq x \leq c - \epsilon$, where $p(x)$ is negative, solutions of (1) are asymptotically of the form

$$(3) \quad c_3 [-p(x)]^{-1/4} \exp\{\lambda \int [-p(x)]^{1/2} dx\} + c_4 [-p(x)]^{-1/4} \exp\{-\lambda \int [-p(x)]^{1/2} dx\}.$$

The validity of these asymptotic forms depends on the fact that $\rho(\xi)$ in 4.1(4) is a bounded function, and hence it is clear that neither of the asymptotic forms can be valid at $x = c$. To the right of $x = c$, (2) shows that every solution of (1) has an *oscillatory* character; and to the left of $x = c$, (3) shows that every solution of (2) has a *monotonic* character. In the immediate vicinity of $x = c$, the *transition* takes place from one type of behavior to the other.

In this situation two problems arise. The first of these is the problem of finding the connection between the constants c_1 and c_2 on the one hand, and the constants c_3 and c_4 on the other hand, if (2) and (3) represent asymptotically the same solution of (1) in different intervals; and the second problem is the determination of the asymptotic form of the solutions of (1) in the interval $(c - \epsilon, c + \epsilon)$.

There are in essence two ingenious methods for solving the first problem. The first of these methods was used by Jeffreys in 1923, and rediscovered by Kramers a few years later. It is based on the remark that sufficiently near to $x = c$, $p(x)$ may be approximated by the linear function $(x - c)\, p'(c)$, and $r(x)$ may be neglected. The resulting differential equation can then be solved in terms of Bessel functions of order $\pm 1/3$. A comparison of the asymptotic forms of Bessel functions with (2) and (3) leads to the desired *connection formulas* between c_1 and c_2 on the one hand, and c_3 and c_4 on the other hand. The second method was developed by Zwaan in 1929, and it avoids the transition point altogether. If $p(x)$ and $r(x)$ are analytic functions of x, the differential equation is integrated along a path in the complex plane which consists of the real intervals $(a, c - \epsilon)$ and $(c + \epsilon, b)$, and of a semi-circle in the complex plane joining $c - \epsilon$ and $c + \epsilon$. Along this path, ρ of 4.1(4) is bounded and Liouville's method (or a variant of it) can be applied. This method leads to the same connection formulas as the first one. Both methods can be extended to cases when $p(x)$ has a zero of an arbitrary order. They are known as the WKB method, or sometimes also the WKBJ method. For a fuller description see, for instance, Morse and Feshbach (1953, p. 1092 ff.).

The second problem, the determination of asymptotic forms of the solutions of (1) in the interval $(c - \epsilon, c + \epsilon)$, is much more difficult. There seems to be no simple elementary function which describes the transition from monotonic to oscillatory behavior, and it seems plausible that the asymptotic forms will involve some higher transcendental function. Now, the simplest differential equation of the form (1) and having a transition point is the equation

$$\frac{d^2 y}{dx^2} + \lambda^2 xy = 0; \tag{4}$$

the solutions of this equation are fairly well known, and it seems tempting to seek asymptotic forms of the solutions of (1) in terms of solutions of (4). Liouville's transformation 4.1 (2) transforms 4.1 (1) into a differential equation with approximately constant coefficients, and similarly, we can find a transformation

$$\xi = \phi(x), \qquad \eta = \psi(x) y \tag{5}$$

which transforms (1) into an equation approximately of the form (4). The transformation (5) carries (1) into

$$\frac{d^2\eta}{d\xi^2} + \frac{1}{\phi'}\left(\frac{\phi''}{\phi'} - 2\frac{\psi'}{\psi}\right)\frac{d\eta}{d\xi} + \left[\frac{\lambda^2 p + r}{\phi'^2} + \frac{\psi}{\phi'^2}\frac{d^2\psi^{-1}}{dx^2}\right]\eta = 0.$$

In order to reduce this differential equation approximately to the form (4), we first determine ψ so that

$$\frac{\phi''}{\phi'} - 2\frac{\psi'}{\psi} = 0, \qquad \psi = \phi'^{1/2}, \tag{6}$$

and then ϕ so that

$$\frac{p}{\phi'^2} = \phi, \qquad \phi\phi'^2 = p. \tag{7}$$

With ϕ and ψ so determined, the differential equation becomes

$$\frac{d^2\eta}{d\xi^2} + \lambda^2 \xi\eta = \rho(\xi)\eta, \tag{8}$$

where

$$(9)\quad \rho(\xi)=\frac{1}{2}\frac{\phi'''}{\phi'^3}-\frac{3}{4}\frac{\phi''^2}{\phi'^4}-\frac{r}{\phi'^2}\,.$$

Under the assumptions on p and r made at the beginning of this section, there is a unique three times continuously differentiable real function ϕ which satisfies (7). For this function, ϕ' is bounded away from zero, $\rho(\xi)$ is a bounded function, and we shall expect that the asymptotic form of the solutions of (8) is

$$(10)\quad c_1 H_1(\xi)+c_2 H_2(\xi),$$

where $H_1(x)$ and $H_2(x)$ are two linearly independent solutions of (4).

This generalization of Liouville's method was originally developed for the purpose of obtaining asymptotic forms of the solutions of (1) in the interval $(c-\epsilon,\ c+\epsilon)$, but it is clear that it can be extended to the entire interval $(a,\ b)$. The extension enables us to dispense with three different asymptotic forms in $(a,\ c-\epsilon)$, $(c-\epsilon,\ c+\epsilon)$, $(c+\epsilon,\ b)$ respectively, and yields a single *uniform asymptotic representation* of the solutions of (1) in $a\le x\le b$. The method was originated by Langer, who developed it in a number of memoirs of which a few are listed at the end of this chapter. Among those who developed further Langer's method we mention in particular Cherry. A survey of the literature regarding this method is available. (See reference at the end of this chapter.)

Before describing this method in greater detail we shall list briefly some properties of the solutions of (4).

4.6. Airy functions

The differential equation

$$(1)\quad \frac{d^2w}{dz^2}-zw=0$$

can be reduced to the differential equation satisfied by Bessel functions of order 1/3 (Watson, 1922, sec. 6.4). We shall use the notation

$$(2)\quad \zeta=\frac{2}{3}z^{3/2},\qquad \omega=e^{2\pi i/3}$$

Two linearly independent solutions of (1) are the so-called *Airy functions* of the first and second kind,

$$(3)\quad Ai(z) = \frac{1}{3} z^{1/2} [I_{-1/3}(\zeta) - I_{1/3}(\zeta)] = \frac{1}{\pi}\left(\frac{z}{3}\right)^{1/2} K_{1/3}(\zeta)$$

$$Bi(z) = \left(\frac{z}{3}\right)^{1/2} [I_{-1/3}(\zeta) + I_{1/3}(\zeta)],$$

and by direct computation

$$(4)\quad Ai(z)\, Bi'(z) - Ai'(z)\, Bi(z) = \pi^{-1}.$$

We also have

$$(5)\quad Ai(-z) = \frac{1}{3} z^{1/2} [J_{-1/3}(\zeta) + J_{1/3}(\zeta)]$$

$$Bi(-z) = \left(\frac{z}{3}\right)^{1/2} [J_{-1/3}(\zeta) - J_{1/3}(\zeta)].$$

For real x, the integral representations

$$(6)\quad Ai(x) = \pi^{-1} \int_0^\infty \cos\left(\frac{t^3}{3} + xt\right) dt$$

$$Bi(x) = \pi^{-1} \int_0^\infty \left[\exp\left(-\frac{t^3}{3} + xt\right) + \sin\left(\frac{t^3}{3} + xt\right)\right] dt$$

hold, and these integrals may be converted into contour integrals which remain valid when x becomes complex.

$Ai(z)$ and $Bi(z)$ are both entire functions of z: they are real for real z. $Ai(z)$ has a string of zeros on the negative real axis, and it has no zeros elsewhere (Watson, 1922, sec. 15.7). For any integer m, $w_m(z) = Ai(\omega^m z)$ is also a solution of (1). By direct computation,

$$(7)\quad w_m(z)\, w_k'(z) - w_m'(z)\, w_k(z) = \frac{1}{\pi\sqrt{3}} \sin\frac{\pi(m-k)}{3} \exp\left[i\pi\left(\frac{m+k}{3} + \frac{1}{2}\right)\right]$$

and hence it is seen that w_m and w_{m+1} are linearly independent. On the other hand, any three of the w_m are linearly dependent, in particular,

(8) $$w_m(z) + \omega\, w_{m+1}(z) + \omega^2 w_{m+2}(z) = 0.$$

Also

(9) $$Bi(z) = i\,[\omega^2 Ai(\omega^2 z) - \omega\, Ai(\omega z)].$$

The power series expansions of $Ai(z)$ and $Bi(z)$ follow from (3). In particular

(10) $$3^{2/3} Ai(0) = 3^{1/6} Bi(0) = \frac{1}{\Gamma(2/3)}$$

$$-3^{1/3} Ai'(0) = 3^{-1/6} Bi'(0) = \frac{1}{\Gamma(1/3)}.$$

The asymptotic behavior of $Ai(z)$ has been investigated in sec. 2.6 for $-\pi/3 < \arg z < \pi/3$. The analysis can be extended by rotating the path of integration in the t-plane of 2.6(3).

(11) $$Ai(z) = \frac{1}{2}\pi^{-\frac{1}{2}} z^{-\frac{1}{4}} e^{-\zeta}[1 + O(\zeta^{-1})]$$

$$\text{as} \quad z \to \infty, \ -\pi < \arg z < \pi.$$

Results valid in sectors containing the negative real axis may be obtained by using (8).

(12) $$Ai(z) = \frac{1}{2}\pi^{-\frac{1}{2}} z^{-\frac{1}{4}} \{e^{-\zeta}[1 + O(\zeta^{-1})] + ie^{\zeta}[1 + O(\zeta^{-1})]\}$$

$$\text{as} \quad z \to \infty, \ \pi/3 < \arg z < 5\pi/3$$

$$Ai(z) = \frac{1}{2}\pi^{-\frac{1}{2}} z^{-\frac{1}{4}} \{e^{-\zeta}[1 + O(\zeta^{-1})] - ie^{\zeta}[1 + O(\zeta^{-1})]\}$$

$$\text{as} \quad z \to \infty, \ -5\pi/3 < \arg z < -\pi/3.$$

From (9),

$$(13)\quad Bi(z) = \pi^{-1/2} z^{-1/4} e^{\zeta} [1 + O(\zeta^{-1})]$$

$$\text{as} \quad z \to \infty, \; -\pi/3 < \arg z < \pi/3.$$

All these asymptotic representations hold uniformly in arg z if z is restricted to a closed sector inside the open sector stated above.

It follows from the asymptotic formulas that the functions

$$(14)\quad (1 + |z|^{1/4})\, e^{\zeta}\, Ai(z) \qquad -\pi \leq \arg z \leq \pi$$

$$(15)\quad (1 + |z|^{1/4})\, e^{-\zeta}\, Bi(z) \qquad -\pi/3 \leq \arg z \leq \pi/3$$

are bounded functions of z. The reciprocals of these functions are also bounded provided that the zeros of the first function are suitably excluded, for instance, if $-\pi + \epsilon \leq \arg z \leq \pi - \epsilon$, $\epsilon > 0$, in (14): the second function has no zeros.

For further information on Airy functions, and for numerical tables of these functions, see Miller (1946).

We conclude this section by proving the following inequality. *If* $w(z)$ *is that solution of* (1) *which satisfies* $w(t) = 0$, $w'(t) = 1$, *then*

$$(16)\quad 0 < \frac{2(xt)^{1/4}\, w(x)}{e^{\xi - \tau} - e^{\tau - \xi}} \leq 1$$

for

$$x > 0, \quad t > 0, \quad \xi = \frac{2}{3} x^{3/2} > 0, \quad \tau = \frac{2}{3} t^{3/2} > 0.$$

The proof depends on a comparison theorem of Sturm's type.

$$v(x) = \frac{e^{\xi - \tau} - e^{\tau - \xi}}{2(xt)^{1/4}}$$

satisfies the differential equation

$$(17)\quad v'' - \left(x + \frac{5}{16x^2}\right) v = 0,$$

and the same initial conditions at t as w. From (1) and (17) it follows that both $v(x)$ and $w(x)$ are $\neq 0$ if $x > 0$, $t > 0$, $x \neq t$. Consider the function $f(x) = w(x)/v(x)$, $x \neq t$. Clearly, $f(x) \to 1$ as $x \to t$, so we put $f(t) = 1$. Also, v and w have the same sign when $x \neq t$.

$$\frac{d}{dx}(w'v - wv') = w''v - wv'' = -\frac{5vw}{16x^2} \leq 0$$

and $w'v - wv'$ is a decreasing function of x. Since this function vanishes at $x = t$, it follows that $f'(x) < 0$ if $x > t$, and $f'(x) > 0$ when $0 < x < t$. Thus, $f'(x)$ has a maximum at $x = t$, and $0 < f(x) \leq 1$ for $0 < x, t < \infty$.

4.7. Asymptotic solutions valid in the transition region

We assume that (a, b) is a bounded interval, $a < c < b$, $p(x)$ is real and and twice continuously differentiable for $a \leq x \leq b$, $p(x) < 0$ for $a \leq x < c$, $p(x) > 0$ for $c < x \leq b$, $p(c) = 0$, $p'(c) \neq 0$, $r(x, \lambda)$ is a bounded function of x and λ when $a \leq x \leq b$ and λ varies over a sectorial domain S: $|\lambda| \geq \lambda_1$, $\phi_0 \leq \arg\lambda \leq \phi_1$ and for each fixed λ in S, $r(x, \lambda)$ is a continuous function of x for $a \leq x \leq b$. We then consider the differential equation

$$(1) \quad y'' + [\lambda^2 p(x) + r(x, \lambda)]\, y = 0,$$

and introduce $\phi(x)$ as the (unique) continuously differentiable real solution of 4.5(7), i.e.

$$(2) \quad \frac{2}{3}[\phi(x)]^{3/2} = f(x) = \int_c^x [p(t)]^{1/2}\, dt \qquad x \geq c$$

$$\frac{2}{3}[-\phi(x)]^{3/2} = f(x) = \int_x^c [-p(t)]^{1/2}\, dt \qquad x \leq c.$$

Here $\phi(x) > 0$ when $x > c$, $\phi(x) < 0$ when $x < c$, and all fractional powers have their positive values.

The functions

$$(3) \quad Y_m(x) = [\phi'(x)]^{-1/2}\, \mathrm{Ai}[-\lambda^{2/3} e^{2\pi im/3}\phi(x)] \qquad m = 0, \pm 1$$

$$Y_2(x) = [\phi'(x)]^{-1/2}\, \mathrm{Bi}[-\lambda^{2/3}\phi(x)]$$

satisfy the differential equation

$$(4) \quad Y''(x) + \left[\lambda^2 p(x) + \frac{1}{2} \{\phi, x\} \right] Y = 0,$$

where

$$\{\phi, x\} = \frac{\phi'''}{\phi'} - \frac{3}{2} \left(\frac{\phi''}{\phi'} \right)^2$$

is the *Schwarzian derivative* of ϕ. This differential equation follows from 4.6(1) and 4.5(7). Since the differential equations (1) and (4) differ in terms which are comparatively small, we consider the Y's as leading terms of formal solutions.

Under the above assumptions, the *differential equation* (1) *possesses solutions which are represented asymptotically, in appropriate sectors of the complex* λ*-plane, by the* Y_m, $m = -1, 0, 1, 2$. The proof is similar to the proofs given in sections 3.3, 3.4, and 4.3; and it will be conducted in several steps.

Let $Y(x)$ *be any solution of* (4), *let* $K(x, t)$ *be that solution of* (4) *which satisfies the initial conditions*

$$(5) \quad K(t, t) = 0, \quad \frac{\partial K}{\partial x}(t, t) = 1,$$

and let $a \le x_0 \le b$; *then the solution of the integral equation*

$$(6) \quad y(x) = Y(x) + \int_{x_0}^{x} K(x, t) F(t, \lambda) y(t)\, dt,$$

where

$$(7) \quad F(x, \lambda) = \frac{1}{2} \{\phi, x\} - r(x, \lambda),$$

satisfies (1).

Proof: Equation (6) is a Volterra integral equation with a continuous kernel. For any fixed value of λ in S, the existence and uniqueness of the solution follows from the general theory of Volterra integral equations (or can be established as in sec. 4.3). This solution is twice continuously differentiable, and substitution in (1) shows that y satisfies the differential equation (1).

For $a \le x,\ t \le b$ and all λ in S,

(8) $$|K(x, t)\, F(t, \lambda)|$$

$$\le \frac{A \exp\left|\frac{2}{3} \operatorname{Re}\{[-\lambda^{2/3}\phi(x)]^{3/2} - [-\lambda^{2/3}\phi(t)]^{3/2}\}\right|}{|\lambda|\,(|\lambda|^{-1/6} + |\phi(x)|^{1/4})\,(|\lambda|^{-1/6} + |\phi(t)|^{1/4})\,[\phi'(x)\,\phi'(t)]^{1/2}}$$

where A is independent of x, t, λ. All fractional powers have their principal values.

The proof of (8) is different according to the position of x and t relative to c and relative to each other, and according to arg λ. We shall give the details for $a \le x,\ t \le c,\ -\pi/2 \le \arg\lambda \le \pi/2$. Since

$$Y_0' Y_2 - Y_0 Y_2' = -\lambda^{2/3}(Ai'Bi - Ai\,Bi') = \pi^{-1}\lambda^{2/3},$$

by (3) and 4.6(4), we see that

(9) $$K(x, t) = \pi\lambda^{-2/3}[Y_0(x)\,Y_2(t) - Y_0(t)\,Y_2(x)]$$

satisfies (4) and (5). Now, $\phi < 0$, and $|\arg(-\lambda^{2/3}\phi)| \le \pi/3$, so that the expressions 4.6(14) and 4.6(15) are bounded, and there is a B independent of x, t, and λ such that

$$|Y_0| \le \frac{B \exp\{-\frac{2}{3}\operatorname{Re}[-\lambda^{2/3}\phi]^{3/2}\}}{(1 + |\lambda|^{1/6}|\phi|^{1/4})\,[\phi']^{1/2}},$$

with a similar estimate for Y_2. Applying these estimates in (9), we prove (8) in this case.

In order to prove (8) in any other case, express Y_0 and Y_2, and hence K, in terms of two linearly independent solutions of (4), of which one is bounded as $\lambda \to \infty$ (this was the solution Y_0 in the above case), and then apply estimates derived from 4.6(14), (15).

We shall now establish the existence of solutions y_m, $m = -1, 0, 1, 2$, corresponding to, and asymptotically represented by, the Y_m. It will be necessary to impose, in each case, restrictions on arg λ. These restrictions correspond to the introduction of branch-cuts in the λ^2-plane.

For y_0 and y_2 we assume Re $\lambda \ge 0$, and define these functions as solutions of the Volterra integral equations

(10) $$y_0(x) = Y_0(x) + \int_a^x K(x, t)\, F(t, \lambda)\, y_0(t)\, dt$$

(11) $y_2(x) = Y_2(x) + \int_c^x K(x, t)\, F(t, \lambda)\, y_2(t)\, dt.$

For any fixed λ in S, the existence of the solutions of these integral equations follows from the general theory of Volterra integral equations (or by successive approximations). We shall prove that $y_0 \sim Y_0$, $y_2 \sim Y_2$ as $\lambda \to \infty$, $\operatorname{Re} \lambda \geq 0$.

If $\operatorname{Re} \lambda \geq 0$, $\operatorname{Im} \lambda \neq 0$, and $a \leq x \leq b$, or if $\lambda \geq 0$ and $a \leq x \leq c$, we have $Y_0 \neq 0$, $Y_2 \neq 0$. In this case we may put $y_0 = Y_0 z_0$, $y_2 = Y_2 z_2$, and obtain integral equations for z_0 and z_2. Since under these circumstances 4.6(14) and 4.6(15) are bounded, and bounded away from zero, we have

$$\left| \frac{Y_0(t)}{Y_0(x)} \right| \leq B \left[\frac{\phi'(x)}{\phi'(t)} \right]^{1/2} \frac{|\lambda|^{-1/6} + |\phi(x)|^{1/4}}{|\lambda|^{-1/6} + |\phi(t)|^{1/4}}$$

$$\times \exp\left(-\frac{2}{3} \operatorname{Re}\{[-\lambda^{2/3} \phi(t)]^{3/2} - [-\lambda^{2/3} \phi(x)]^{3/2}\} \right),$$

and a similar estimate for $|Y_2(t)/Y_2(x)|$. Combining these estimates with (8), we obtain inequalities,

$$(12) \qquad \left| \frac{Y_0(t)}{Y_0(x)} K(x, t)\, F(t, \lambda) \right| \leq \frac{C}{|\lambda|\, \phi'(t)\, |\phi(t)|^{1/2}}$$

$a \leq t \leq x \leq b$ and $\operatorname{Im} \lambda \neq 0$, or $a \leq t \leq x \leq c$ and $\lambda > 0$.

$$(13) \qquad \left| \frac{Y_2(t)}{Y_2(x)} K(x, t)\, F(t, \lambda) \right| \leq \frac{C}{|\lambda|\, \phi'(t)\, |\phi(t)|^{1/2}}$$

$a \leq x \leq b$, $|t - c| \leq |x - c|$, and $\operatorname{Im} \lambda \neq 0$,
or $a \leq x \leq t \leq c$ and $\lambda > 0$

for the kernels of the integral equations for z_0 and z_2. For each fixed λ, the integral equation for z_m, $m = 0, 2$ has a bounded solution. Let $Z_m(\lambda)$ be the maximum of $|z_m(x)|$ for $a \leq x \leq b$. From the integral equation and (12), (13),

$$Z_m(\lambda) \leq 1 + \frac{c}{\lambda} Z_m(\lambda) \int_a^b \frac{dt}{|p(t)|^{1/2}}.$$

For sufficiently large $|\lambda|$, $|Z_m(\lambda)| \leq 2$, and from (10), (11), (12), (13),

(14) $y_m(x) = Y_m(x)\,[1 + O(\lambda^{-1})]$, uniformly in x, as $\lambda \to \infty$, $m = 0, 2$

provided that Re $\lambda \geq 0$, $a \leq x \leq b$, and also that in case of a real λ, x is restricted to $a \leq x \leq c$.

In the case of a positive real λ, Y_0 and Y_2 have zeros when $x > c$, and (14) fails to hold near such zeros. An appropriately modified result can be derived from (10) and (11) by using (9) and estimates of $Y_0(t)$, $Y_2(t)$. From (9)

$$y_0(x) = Y_0(x)\,[1 + \pi\lambda^{-2/3} \int_a^x Y_2(t)\,F(t, \lambda)\,y_0(t)\,dt]$$

$$- \pi\lambda^{-2/3}\,Y_2(x) \int_a^x Y_0(t)\,F(t, \lambda)\,y_0(t)\,dt.$$

Here the integrals can be broken up,

$$\int_a^x = \int_a^c + \int_c^x,$$

(14) may be used in the first integral, and thus it can be proved

(15) $y_0(x) = Y_0(x)\,[1 + O(\lambda^{-1})] + Y_2(x)\,O(\lambda^{-1})$,

uniformly in x, as $\lambda \to \infty$, $c \leq x \leq b$, $\lambda > 0$.

Similar is the proof of

(16) $y_2(x) = Y_2(x)\,[1 + O(\lambda^{-1})] + Y_0(x)\,O(\lambda^{-1})$,

uniformly in x, as $\lambda \to \infty$, $c \leq x \leq b$, $\lambda > 0$.

Equations (14) to (16) describe completely the asymptotic properties of y_0 and y_2.

The work on $y_{\pm 1}$ is similar. We define y_1 and y_{-1} by the integral equations

(17) $y_m(x) = Y_m(x) - \int_x^b K(x, t)\,F(t, \lambda)\,y_m(t)\,dt$ $\qquad m = 1, -1,$

and analyze the solutions of these equations in a manner similar to the analysis of (10), (11), assuming Im $\lambda \geq 0$ in the case of y_1, and Im $\lambda \leq 0$ in the case of y_{-1}. Y_1 and Y_{-1} have zeros when λ is imaginary and $x < c$, and in this case the asymptotic forms need some modification. The final results are

(18) $y_1(x) = Y_1(x)\,[1 + O(\lambda^{-1})]$, uniformly in x, as $\lambda \to \infty$

$$a \le x \le b, \quad \operatorname{Im} \lambda \ge 0, \quad \operatorname{Re} \lambda \ne 0 \quad \text{or} \quad c \le x \le b, \quad -i\lambda > 0$$

(19) $y_1(x) = Y_1(x)\,[1 + O(\lambda^{-1})] + Y_{-1}(x)\,O(\lambda^{-1})$,

$$\text{uniformly in } x, \quad \text{as} \quad \lambda \to \infty, \quad a \le x \le c, \quad -i\lambda > 0$$

(20) $y_{-1}(x) = Y_{-1}(x)\,[1 + O(\lambda^{-1})]$, uniformly in x, as $\lambda \to \infty$

$$a \le x \le b, \quad \operatorname{Im} \lambda \le 0, \quad \operatorname{Re} \lambda \ne 0, \quad \text{or} \quad c \le x \le b, \quad i\lambda > 0$$

(21) $y_{-1}(x) = Y_{-1}(x)\,[1 + O(\lambda^{-1})] + Y_1(x)\,O(\lambda^{-1})$,

$$\text{uniformly in } x, \quad \text{as} \quad \lambda \to \infty, \quad a \le x \le c, \quad i\lambda > 0.$$

Equations (14) to (16) and (18) to (21) establish the result stated at the beginning of this section. By a more careful discussion of the integral equations it is possible to show that also $y'_m \sim Y'_m$.

If $r(x, \lambda)$ can be expanded in powers of λ^{-1}, then it is possible to develop formal solutions of (1). The approximations discussed in this section appear as the leading terms of the formal solutions. As in the case discussed in the earlier parts of this chapter, there are two kinds of formal solutions. The first kind corresponds to 4.2(3), and it is of the form

(22) $$Y(x) \sum_{n=0}^{\infty} \alpha_n(x)\,\lambda^{-n} + Y'(x) \sum_{n=0}^{\infty} \beta_n(x)\,\lambda^{-n-1},$$

where Y is a solution of (4), and the $\alpha_n(x)$ and $\beta_n(x)$ are functions of x which are independent of λ. Recurrent differential equations for these functions may be obtained by substituting (22) in (1), and equating coefficients of like powers of λ. This approach was used by Langer (1949).

A second kind of formal solution corresponds to 4.2(9). This solution is of the form (3), except that instead of $\phi(x)$ we have a function $\phi(x, \lambda)$ which depends on λ, and possesses a formal expansion

(23) $$\sum_{n=0}^{\infty} \phi_n(x)\,\lambda^{-n},$$

where $\phi_0(x)$ is the differentiable real solution of 4.5(7), and recurrent differential equations for $\phi_1, \phi_2, \ldots$ are obtained by substituting

(24) $\phi'^{-1/2}\, Ai\,(-\lambda^{2/3}\,\phi)$

in (1), and equating coefficients of powers of λ. Such solutions were used by Cherry (1950).

The differential equation

$$y'' + q(x, \lambda)\, y = 0, \qquad q(x, \lambda) = \sum_{n=0}^{\infty} q_n(x)\, \lambda^{2-n}$$

is more general that (1), because it contains the term $q_1(x)\lambda$. This term causes certain additional complications which were also discussed by Langer (1949).

The extension of these results to the case of a complex variable was undertaken both by Langer (1932) and Cherry (1950); the extension to unbounded x was investigated by Cherry (1950).

4.8. Uniform asymptotic representations of Bessel functions

We conclude with an application of the results of the preceding section to the differential equation 4.4 (3), thereby obtaining an asymptotic representation of $J_\lambda(\lambda x)$ which holds, uniformly for all positive x, as $\lambda \to \infty$, Re $\lambda \geq 0$. The result to be obtained includes the results of sec. 4.4 as far as they relate to $J_\lambda(\lambda x)$, and in addition fills the gap, $b < x < a$ $(b < 1 < a)$ left by the work of sec. 4.4.

Let us, then, apply the methods of the preceding section to the differential equation 4.4 (3), where

$$(1) \quad p(x) = 1 - \frac{1}{x^2}, \qquad r(x, \lambda) = \frac{1}{4x^2}.$$

The transition point is at $x = 1$. The function ϕ is determined by the differential equation

$$(2) \quad \phi\phi'^2 = 1 - \frac{1}{x^2},$$

and 4.7 (2) becomes

$$(3) \quad \frac{2}{3}\,[-\phi(x)]^{3/2} = \int_x^1 (t^{-2} - 1)^{1/2}\, dt$$

$$= -(1 - x^2)^{1/2} + \log \frac{1 + (1 - x^2)^{1/2}}{x} = -\beta(x) \qquad 0 < x \leq 1$$

$$(4) \quad \frac{2}{3}\,[\phi(x)]^{3/2} = \int_1^x (1-t^{-2})^{1/2}\,dt = (x^2-1)^{1/2} - \cos^{-1} x^{-1} = f(x)$$

$$1 \leq x < \infty,$$

see also 4.4 (4) and (14).

Since $\phi(x)$ is an analytic function, and $\phi'(x) \neq 0$, clearly

$$F(x) = \frac{1}{2}\{\phi, x\} - r(x, \lambda) = \frac{1}{2}\frac{\phi'''}{\phi'} - \frac{3}{4}\left(\frac{\phi''}{\phi'}\right)^2 - \frac{1}{4x^2}$$

is a continuous function of x for $0 < x < \infty$. To investigate the behavior of $F(x)$ as $x \to 0$ or $x \to \infty$, we use the chain rule

$$\{\phi, x\} = \{\phi, u\}\left(\frac{du}{dx}\right)^2 + \{u, x\}$$

for the Schwarzian derivative, with $u = \beta$ when $x < 1$, and $u = f$ when $x > 1$. By a simple computation

$$F(x) = \frac{5\beta'^2}{18\beta^2} + \frac{1}{2}\{\beta, x\} - \frac{1}{4x^2} = \frac{5\beta'^2}{18\beta^2} - \frac{4+x^2}{4(1-x^2)^2} \qquad 0 < x < 1$$

$$F(x) = \frac{5f'^2}{18f^2} - \frac{4+x^2}{4(1-x^2)^2} \qquad 1 < x < \infty.$$

From (3) and (4),

$$\beta = O(\log x), \qquad \beta' = O(x^{-1}) \qquad \text{as} \quad x \to 0$$

$$f = O(x), \qquad f' = O(1) \qquad \text{as} \quad x \to \infty,$$

and it follows that

$$(5) \quad F(x) = O[(x \log x)^{-2}] \qquad \text{as} \quad x \to 0$$

$$F(x) = O(x^{-2}) \qquad \text{as} \quad x \to \infty.$$

Furthermore, as in sec. 4.7,

$$(6) \quad \left|\frac{Y_0(t)}{Y_0(x)}\,K(x, t)\right| \leq \frac{C}{|\lambda|\,\phi'(t)\,|\phi(t)|^{1/2}} = \frac{C}{|\lambda|}\left|\frac{t^2}{1-t^2}\right|^{1/2}$$

except near zeros of $Y_0(x)$.

We thus see that we may take $x = 0$ as the fixed limit in the integral equation for $z_0(x)$. For the kernel we have estimates from (5) and (6) which show that the kernel is integrable in $(0, \infty)$. The method of sec. 4.7 may be applied to our problem in spite of the singular point at $x = 0$ of the differential equation. Equation 4.7 (14) with $m = 0$ holds, and since it follows from (5) and (6) that

$$\int_0^x \left| \frac{Y_0(t)}{Y_0(x)} K(x, t) F(t) \right| dt = 0\left(\int_0^x \frac{dt}{t(\log t)^2} \right) = O\left(\frac{1}{\log x} \right) \qquad \text{as} \quad x \to 0,$$

we see that for small x, the term $O(\lambda^{-1})$ in 4.7 (14) may be strengthened to

$$(7) \quad O\left(\frac{1}{\lambda \log x} \right).$$

We then conclude that the differential equation 4.4 (3) possesses a solution $y_0(x)$ for which

$$(8) \quad y_0(x) = \phi'^{-1/2} Ai(-\lambda^{2/3} \phi) [1 + O(\lambda^{-1})],$$

holds for $0 < x < \infty$, $\lambda \to \infty$, $\operatorname{Re} \lambda \geq 0$, except that in the case of a positive real λ, and $x > 1$, the error term needs some modification. For $0 < x \leq b < 1$, the error term may be strengthened to (7).

In order to identify y_0 in terms of Bessel functions, let us take $0 < x \leq b < 1$ in (8), so that $\phi(x) \leq \phi(b) < 0$, the Airy function may be replaced by its asymptotic representation 4.6 (11), the O-term may be strengthened to (7), and we obtain

$$y_0(x) = \frac{1}{2} \pi^{-1/2} \lambda^{-1/6} \phi'^{-1/2} (-\phi)^{1/4} e^{\lambda\beta} \left[1 + O\left(\frac{1}{\lambda \log x} \right) \right]$$

$$= \frac{1}{2} \pi^{-1/2} \lambda^{-1/6} a e^{\lambda\beta} \left[1 + O\left(\frac{1}{\lambda \log x} \right) \right] \qquad 0 < x \leq b < 1,$$

where a is the function defined in 4.4 (5). A comparison with 4.4 (6) and 4.4 (13) now shows that

$$J_\lambda(\lambda x) = \frac{2\pi^{1/2}}{\Gamma(\lambda + 1)} e^{-\lambda} \lambda^{\lambda + 1/6} x^{-1/2} y_0(x) \qquad 0 < x < \infty$$

and hence

$$(9)\quad J_\lambda(\lambda x) = \frac{2\pi^{1/2}}{\Gamma(\lambda+1)} e^{-\lambda} \lambda^{\lambda+1/6} (x\phi')^{-1/2} Ai(-\lambda^{2/3}\phi)[1 + O(\lambda^{-1})],$$

with the same remarks about the error term as in (8). By applying Stirling's formula to $\Gamma(\lambda + 1)$, this result may be put in the simpler, if weaker form

$$(10)\quad J_\lambda(\lambda x) = \left(\frac{1}{2}\lambda^{2/3} x \phi'\right)^{-1/2} Ai(-\lambda^{2/3}\phi)[1 + O(\lambda^{-1})],$$

uniformly in x, $0 < x < \infty$, as $\lambda \to \infty$, Re $\lambda \geq 0$, except that the error term needs some modification near zeros of $Ai(-\lambda^{2/3}\phi)$. Note that in (10) the error term contains the error of Stirling's formula and cannot be strengthened for small x.

In the process of deriving (7) we have seen that our present result includes 4.4(1). Let us show that (7) also includes the sum of the two equations 4.4(2). To do this, we assume $x \geq a > 1$, $\phi(x) \geq \phi(a) > 0$, and apply 4.6(12) to show that

$$J_\lambda(\lambda x) \sim \left(\frac{1}{2}\pi\lambda x\right)^{-1/2} \phi'^{-1/2}.\phi^{-1/4} \cos[\lambda f(x) - \pi/4] \qquad 1 < a \leq x < \infty$$

or

$$J_\lambda(\lambda x) \sim \left(\frac{1}{2}\pi\lambda x\right)^{-1/2} a(x) \cos[\lambda f(x) - \pi/4] \qquad 1 < a \leq x < \infty,$$

and this is in agreement with 4.4(2).

The main result of this section, (6), has been extended to complex values of x, and approximations of higher order have been obtained by Cherry (1948).

REFERENCES

Birkhoff, G.D., 1908: *Trans. Amer. Math. Soc.* 9, 219-231 and 380-382.

Cherry, T.M., 1948: *J. London Math. Soc.* 24, 121-130.

Cherry, T.M., 1950: *Trans. Amer. Math. Soc.* 68, 224-257.

Horn, Jakob, 1899: *Math. Ann.* 52, 271-292 and 340-362.

Ince, E.L., 1927: *Ordinary differential equations*, Longmans, Green and Co.

Jeffreys, Harold, 1923: *Proc. London Math. Soc.* (2) 23, 428-436.

Jeffreys, Harold, 1953: *Proc. Cambridge Philos. Soc.* 49, 601-611.

Kamke, E.W.H., 1930: *Differentialgleichungen reeller Funktionen.* Leipzig.

Kamke, E.W.H., 1944: *Differentialgleichungen. Lösungsmethoden und Lösungen.* Third Edition, Leipzig.

Langer, R.E., 1932: *Trans. Amer. Math. Soc.* 34, 447-480.

Langer, R.E., 1934: *Bull. Amer. Math. Soc.* 40, 545-582.

Langer, R.E., 1935: *Trans. Amer. Math. Soc.* 37, 397-416.

Langer, R.E., 1949: *Trans. Amer. Math. Soc.* 67, 461-490.

Miller, J.C.P., 1946: *The Airy integral.* Cambridge.

Morse, P.M. and Herman Feshbach, 1953: *Methods of theoretical physics*, 2 vols. McGraw-Hill.

Tamarkin, J.D., 1928: *Math. Z.* 27, 1-54.

Trjitzinsky, W.J., 1936: *Acta Math.* 67, 1-50.

Trjitzinsky, W.J., 1938: *Bull. Amer. Math. Soc.* 44, 208-222.

Turrittin, H.L., 1936: *Amer. J. Math.* 58, 364-376.

Turrittin, H.L., 1952: *Contributions to the theory of non-linear oscillations*, v. II. Ann. of Math. Study, no. 29, 81-115.

Wasow, Wolfgang, 1953: *Introduction to the asymptotic theory of ordinary linear differential equations.* Working paper. National Bureau of Standards.

Watson, G.N., 1922: *A treatise on the theory of Bessel functions.* Cambridge.

See also:

Asymptotic solutions of differential equations with turning points. Review of the literature. Technical Report 1, Contract Nonr-220(11). Reference no. NR 043-121. Department of Mathematics, California Institute of Technology, 1953.

A CATALOGUE OF SELECTED DOVER BOOKS IN ALL FIELDS OF INTEREST

A CATALOG OF SELECTED DOVER BOOKS IN ALL FIELDS OF INTEREST

LASERS AND HOLOGRAPHY, Winston E. Kock. Sound introduction to burgeoning field, expanded (1981) for second edition. 84 illustrations. 160pp. 5⅜ × 8¼. (EUK) 24041-X Pa. $3.50

FLORAL STAINED GLASS PATTERN BOOK, Ed Sibbett, Jr. 96 exquisite floral patterns—irises, poppie, lilies, tulips, geometrics, abstracts, etc.—adaptable to innumerable stained glass projects. 64pp. 8¼ × 11. 24259-5 Pa. $3.50

THE HISTORY OF THE LEWIS AND CLARK EXPEDITION, Meriwether Lewis and William Clark. Edited by Eliott Coues. Great classic edition of Lewis and Clark's day-by-day journals. Complete 1893 edition, edited by Eliott Coues from Biddle's authorized 1814 history. 1508pp. 5⅜ × 8½.
21268-8, 21269-6, 21270-X Pa. Three-vol. set $22.50

ORLEY FARM, Anthony Trollope. Three-dimensional tale of great criminal case. Original Millais illustrations illuminate marvelous panorama of Victorian society. Plot was author's favorite. 736pp. 5⅜ × 8½. 24181-5 Pa. $8.95

THE CLAVERINGS, Anthony Trollope. Major novel, chronicling aspects of British Victorian society, personalities. 16 plates by M. Edwards; first reprint of full text. 412pp. 5⅜ × 8½. 23464-9 Pa. $6.00

EINSTEIN'S THEORY OF RELATIVITY, Max Born. Finest semi-technical account; much explanation of ideas and math not readily available elsewhere on this level. 376pp. 5⅜ × 8½. 60769-0 Pa. $5.00

COMPUTABILITY AND UNSOLVABILITY, Martin Davis. Classic graduate-level introduction th theory of computability, usually referred to as theory of recurrent functions. New preface and appendix. 288pp. 5⅜ × 8½. 61471-9 Pa. $6.50

THE GODS OF THE EGYPTIANS, E.A. Wallis Budge. Never excelled for richness, fullness: all gods, goddesses, demons, mythical figures of Ancient Egypt; their legends, rites, incarnations, etc. Over 225 illustrations, plus 6 color plates. 988pp. 6⅛ × 9¼. (EBE) 22055-9, 22056-7 Pa., Two-vol. set $20.00

THE I CHING (THE BOOK OF CHANGES), translated by James Legge. Most penetrating divination manual ever prepared. Indispensable to study of early Oriental civilizations, to modern inquiring reader. 448pp. 5⅜ × 8½.
21062-6 Pa. $6.50

THE CRAFTSMAN'S HANDBOOK, Cennino Cennini. 15th-century handbook, school of Giotto, explains applying gold, silver leaf; gesso; fresco painting, grinding pigments, etc. 142pp. 6⅛ × 9¼. 20054-X Pa. $3.50

AN ATLAS OF ANATOMY FOR ARTISTS, Fritz Schider. Finest text, working book. Full text, plus anatomical illustrations; plates by great artists showing anatomy. 593 illustrations. 192pp. 7⅛ × 10¼. 20241-0 Pa. $6.00

EASY-TO-MAKE STAINED GLASS LIGHTCATCHERS, Ed Sibbett, Jr. 67 designs for most enjoyable ornaments: fruits, birds, teddy bears, trumpet, etc. Full size templates. 64pp. 8¼ × 11. 24081-9 Pa. $3.95

TRIAD OPTICAL ILLUSIONS AND HOW TO DESIGN THEM, Harry Turner. Triad explained in 32 pages of text, with 32 pages of Escher-like patterns on coloring stock. 92 figures. 32 plates. 64pp. 8¼ × 11. 23549-1 Pa. $2.50

25 KITES THAT FLY, Leslie Hunt. Full, easy-to-follow instructions for kites made from inexpensive materials. Many novelties. 70 illustrations. 110pp. 5⅜ × 8½. 22550-X Pa. $1.95

PIANO TUNING, J. Cree Fischer. Clearest, best book for beginner, amateur. Simple repairs, raising dropped notes, tuning by easy method of flattened fifths. No previous skills needed. 4 illustrations. 201pp. 5⅜ × 8½. 23267-0 Pa. $3.50

EARLY AMERICAN IRON-ON TRANSFER PATTERNS, edited by Rita Weiss. 75 designs, borders, alphabets, from traditional American sources. 48pp. 8¼ × 11. 23162-3 Pa. $1.95

CROCHETING EDGINGS, edited by Rita Weiss. Over 100 of the best designs for these lovely trims for a host of household items. Complete instructions, illustrations. 48pp. 8¼ × 11. 24031-2 Pa. $2.00

FINGER PLAYS FOR NURSERY AND KINDERGARTEN, Emilie Poulsson. 18 finger plays with music (voice and piano); entertaining, instructive. Counting, nature lore, etc. Victorian classic. 53 illustrations. 80pp. 6½ × 9¼. 22588-7 Pa. $1.95

BOSTON THEN AND NOW, Peter Vanderwarker. Here in 59 side-by-side views are photographic documentations of the city's past and present. 119 photographs. Full captions. 122pp. 8¼ × 11. 24312-5 Pa. $6.95

CROCHETING BEDSPREADS, edited by Rita Weiss. 22 patterns, originally published in three instruction books 1939-41. 39 photos, 8 charts. Instructions. 48pp. 8¼ × 11. 23610-2 Pa. $2.00

HAWTHORNE ON PAINTING, Charles W. Hawthorne. Collected from notes taken by students at famous Cape Cod School; hundreds of direct, personal *apercus*, ideas, suggestions. 91pp. 5⅜ × 8½. 20653-X Pa. $2.50

THERMODYNAMICS, Enrico Fermi. A classic of modern science. Clear, organized treatment of systems, first and second laws, entropy, thermodynamic potentials, etc. Calculus required. 160pp. 5⅜ × 8½. 60361-X Pa. $4.00

TEN BOOKS ON ARCHITECTURE, Vitruvius. The most important book ever written on architecture. Early Roman aesthetics, technology, classical orders, site selection, all other aspects. Morgan translation. 331pp. 5⅜ × 8½. 20645-9 Pa. $5.50

THE CORNELL BREAD BOOK, Clive M. McCay and Jeanette B. McCay. Famed high-protein recipe incorporated into breads, rolls, buns, coffee cakes, pizza, pie crusts, more. Nearly 50 illustrations. 48pp. 8¼ × 11. 23995-0 Pa. $2.00

THE CRAFTSMAN'S HANDBOOK, Cennino Cennini. 15th-century handbook, school of Giotto, explains applying gold, silver leaf; gesso; fresco painting, grinding pigments, etc. 142pp. 6⅛ × 9¼. 20054-X Pa. $3.50

FRANK LLOYD WRIGHT'S FALLINGWATER, Donald Hoffmann. Full story of Wright's masterwork at Bear Run, Pa. 100 photographs of site, construction, and details of completed structure. 112pp. 9¼ × 10. 23671-4 Pa. $6.50

OVAL STAINED GLASS PATTERN BOOK, C. Eaton. 60 new designs framed in shape of an oval. Greater complexity, challenge with sinuous cats, birds, mandalas framed in antique shape. 64pp. 8¼ × 11. 24519-5 Pa. $3.50

CHILDREN'S BOOKPLATES AND LABELS, Ed Sibbett, Jr. 6 each of 12 types based on *Wizard of Oz, Alice,* nursery rhymes, fairy tales. Perforated; full color. 24pp. 8¼ × 11. 23538-6 Pa. $2.95

READY-TO-USE VICTORIAN COLOR STICKERS: 96 Pressure-Sensitive Seals, Carol Belanger Grafton. Drawn from authentic period sources. Motifs include heads of men, women, children, plus florals, animals, birds, more. Will adhere to any clean surface. 8pp. 8½ × 11. 24551-9 Pa. $2.95

CUT AND FOLD PAPER SPACESHIPS THAT FLY, Michael Grater. 16 colorful, easy-to-build spaceships that really fly. Star Shuttle, Lunar Freighter, Star Probe, 13 others. 32pp. 8¼ × 11. 23978-0 Pa. $2.50

CUT AND ASSEMBLE PAPER AIRPLANES THAT FLY, Arthur Baker. 8 aerodynamically sound, ready-to-build paper airplanes, designed with latest techniques. Fly *Pegasus, Daedalus, Songbird,* 5 other aircraft. Instructions. 32pp. 9¼ × 11¼. 24302-8 Pa. $3.95

SIDELIGHTS ON RELATIVITY, Albert Einstein. Two lectures delivered in 1920-21: *Ether and Relativity* and *Geometry and Experience.* Elegant ideas in non-mathematical form. 56pp. 5⅜ × 8½. 24511-X Pa. $2.25

FADS AND FALLACIES IN THE NAME OF SCIENCE, Martin Gardner. Fair, witty appraisal of cranks and quacks of science: Velikovsky, orgone energy, Bridey Murphy, medical fads, etc. 373pp. 5⅜ × 8½. 20394-8 Pa. $5.50

VACATION HOMES AND CABINS, U.S. Dept. of Agriculture. Complete plans for 16 cabins, vacation homes and other shelters. 105pp. 9 × 12. 23631-5 Pa. $4.50

HOW TO BUILD A WOOD-FRAME HOUSE, L.O. Anderson. Placement, foundations, framing, sheathing, roof, insulation, plaster, finishing—almost everything else. 179 illustrations. 223pp. 7⅞ × 10¾. 22954-8 Pa. $5.50

THE MYSTERY OF A HANSOM CAB, Fergus W. Hume. Bizarre murder in a hansom cab leads to engrossing investigation. Memorable characters, rich atmosphere. 19th-century bestseller, still enjoyable, exciting. 256pp. 5⅜ × 8. 21956-9 Pa. $4.00

MANUAL OF TRADITIONAL WOOD CARVING, edited by Paul N. Hasluck. Possibly the best book in English on the craft of wood carving. Practical instructions, along with 1,146 working drawings and photographic illustrations. 576pp. 6½ × 9¼. 23489-4 Pa. $8.95

WHITTLING AND WOODCARVING, E.J Tangerman. Best book on market; clear, full. If you can cut a potato, you can carve toys, puzzles, chains, etc. Over 464 illustrations. 293pp. 5⅜ × 8½. 20965-2 Pa. $4.95

AMERICAN TRADEMARK DESIGNS, Barbara Baer Capitman. 732 marks, logos and corporate-identity symbols. Categories include entertainment, heavy industry, food and beverage. All black-and-white in standard forms. 160pp. 8⅛ × 11. 23259-X Pa. $6.00

DECORATIVE FRAMES AND BORDERS, edited by Edmund V. Gillon, Jr. Largest collection of borders and frames ever compiled for use of artists and designers. Renaissance, neo-Greek, Art Nouveau, Art Deco, to mention only a few styles. 396 illustrations. 192pp. 8⅜ × 11¼. 22928-9 Pa. $6.00

THE MURDER BOOK OF J.G. REEDER, Edgar Wallace. Eight suspenseful stories by bestselling mystery writer of 20s and 30s. Features the donnish Mr. J.G. Reeder of Public Prosecutor's Office. 128pp. 5⅜ × 8½. (Available in U.S. only) 24374-5 Pa. $3.50

ANNE ORR'S CHARTED DESIGNS, Anne Orr. Best designs by premier needlework designer, all on charts: flowers, borders, birds, children, alphabets, etc. Over 100 charts, 10 in color. Total of 40pp. 8¼ × 11. 23704-4 Pa. $2.25

BASIC CONSTRUCTION TECHNIQUES FOR HOUSES AND SMALL BUILDINGS SIMPLY EXPLAINED, U.S. Bureau of Naval Personnel. Grading, masonry, woodworking, floor and wall framing, roof framing, plastering, tile setting, much more. Over 675 illustrations. 568pp. 6½ × 9¼. 20242-9 Pa. $8.95

MATISSE LINE DRAWINGS AND PRINTS, Henri Matisse. Representative collection of female nudes, faces, still lifes, experimental works, etc., from 1898 to 1948. 50 illustrations. 48pp. 8⅜ × 11¼. 23877-6 Pa. $2.50

HOW TO PLAY THE CHESS OPENINGS, Eugene Znosko-Borovsky. Clear, profound examinations of just what each opening is intended to do and how opponent can counter. Many sample games. 147pp. 5⅜ × 8½. 22795-2 Pa. $2.95

DUPLICATE BRIDGE, Alfred Sheinwold. Clear, thorough, easily followed account: rules, etiquette, scoring, strategy, bidding; Goren's point-count system, Blackwood and Gerber conventions, etc. 158pp. 5⅜ × 8½. 22741-3 Pa. $3.00

SARGENT PORTRAIT DRAWINGS, J.S. Sargent. Collection of 42 portraits reveals technical skill and intuitive eye of noted American portrait painter, John Singer Sargent. 48pp. 8¼ × 11⅛. 24524-1 Pa. $2.95

ENTERTAINING SCIENCE EXPERIMENTS WITH EVERYDAY OBJECTS, Martin Gardner. Over 100 experiments for youngsters. Will amuse, astonish, teach, and entertain. Over 100 illustrations. 127pp. 5⅜ × 8½. 24201-3 Pa. $2.50

TEDDY BEAR PAPER DOLLS IN FULL COLOR: A Family of Four Bears and Their Costumes, Crystal Collins. A family of four Teddy Bear paper dolls and nearly 60 cut-out costumes. Full color, printed one side only. 32pp. 9¼ × 12¼. 24550-0 Pa. $3.50

NEW CALLIGRAPHIC ORNAMENTS AND FLOURISHES, Arthur Baker. Unusual, multi-useable material: arrows, pointing hands, brackets and frames, ovals, swirls, birds, etc. Nearly 700 illustrations. 80pp. 8⅜ × 11¼. 24095-9 Pa. $3.50

DINOSAUR DIORAMAS TO CUT & ASSEMBLE, M. Kalmenoff. Two complete three-dimensional scenes in full color, with 31 cut-out animals and plants. Excellent educational toy for youngsters. Instructions; 2 assembly diagrams. 32pp. 9¼ × 12¼. 24541-1 Pa. $3.95

SILHOUETTES: A PICTORIAL ARCHIVE OF VARIED ILLUSTRATIONS, edited by Carol Belanger Grafton. Over 600 silhouettes from the 18th to 20th centuries. Profiles and full figures of men, women, children, birds, animals, groups and scenes, nature, ships, an alphabet. 144pp. 8⅜ × 11¼. 23781-8 Pa. $4.50

SURREAL STICKERS AND UNREAL STAMPS, William Rowe. 224 haunting, hilarious stamps on gummed, perforated stock, with images of elephants, geisha girls, George Washington, etc. 16pp. one side. 8¼ × 11. 24371-0 Pa. $3.50

GOURMET KITCHEN LABELS, Ed Sibbett, Jr. 112 full-color labels (4 copies each of 28 designs). Fruit, bread, other culinary motifs. Gummed and perforated. 16pp. 8¼ × 11. 24087-8 Pa. $2.95

PATTERNS AND INSTRUCTIONS FOR CARVING AUTHENTIC BIRDS, H.D. Green. Detailed instructions, 27 diagrams, 85 photographs for carving 15 species of birds so life-like, they'll seem ready to fly! 8¼ × 11. 24222-6 Pa. $2.75

FLATLAND, E.A. Abbott. Science-fiction classic explores life of 2-D being in 3-D world. 16 illustrations. 103pp. 5⅜ × 8. 20001-9 Pa. $2.00

DRIED FLOWERS, Sarah Whitlock and Martha Rankin. Concise, clear, practical guide to dehydration, glycerinizing, pressing plant material, and more. Covers use of silica gel. 12 drawings. 32pp. 5⅜ × 8½. 21802-3 Pa. $1.00

EASY-TO-MAKE CANDLES, Gary V. Guy. Learn how easy it is to make all kinds of decorative candles. Step-by-step instructions. 82 illustrations. 48pp. 8¼ × 11. 23881-4 Pa. $2.50

SUPER STICKERS FOR KIDS, Carolyn Bracken. 128 gummed and perforated full-color stickers: GIRL WANTED, KEEP OUT, BORED OF EDUCATION, X-RATED, COMBAT ZONE, many others. 16pp. 8¼ × 11. 24092-4 Pa. $2.50

CUT AND COLOR PAPER MASKS, Michael Grater. Clowns, animals, funny faces...simply color them in, cut them out, and put them together, and you have 9 paper masks to play with and enjoy. 32pp. 8¼ × 11. 23171-2 Pa. $2.25

A CHRISTMAS CAROL: THE ORIGINAL MANUSCRIPT, Charles Dickens. Clear facsimile of Dickens manuscript, on facing pages with final printed text. 8 illustrations by John Leech, 4 in color on covers. 144pp. 8⅜ × 11¼. 20980-6 Pa. $5.95

CARVING SHOREBIRDS, Harry V. Shourds & Anthony Hillman. 16 full-size patterns (all double-page spreads) for 19 North American shorebirds with step-by-step instructions. 72pp. 9¼ × 12¼. 24287-0 Pa. $4.95

THE GENTLE ART OF MATHEMATICS, Dan Pedoe. Mathematical games, probability, the question of infinity, topology, how the laws of algebra work, problems of irrational numbers, and more. 42 figures. 143pp. 5⅜ × 8½. (EBE) 22949-1 Pa. $3.00

READY-TO-USE DOLLHOUSE WALLPAPER, Katzenbach & Warren, Inc. Stripe, 2 floral stripes, 2 allover florals, polka dot; all in full color. 4 sheets (350 sq. in.) of each, enough for average room. 48pp. 8¼ × 11. 23495-9 Pa. $2.95

MINIATURE IRON-ON TRANSFER PATTERNS FOR DOLLHOUSES, DOLLS, AND SMALL PROJECTS, Rita Weiss and Frank Fontana. Over 100 miniature patterns: rugs, bedspreads, quilts, chair seats, etc. In standard dollhouse size. 48pp. 8¼ × 11. 23741-9 Pa. $1.95

THE DINOSAUR COLORING BOOK, Anthony Rao. 45 renderings of dinosaurs, fossil birds, turtles, other creatures of Mesozoic Era. Scientifically accurate. Captions. 48pp. 8¼ × 11. 24022-3 Pa. $2.25

THE BOOK OF WOOD CARVING, Charles Marshall Sayers. Still finest book for beginning student. Fundamentals, technique; gives 34 designs, over 34 projects for panels, bookends, mirrors, etc. 33 photos. 118pp. 7¾ × 10⅝. 23654-4 Pa. $3.95

CARVING COUNTRY CHARACTERS, Bill Higginbotham. Expert advice for beginning, advanced carvers on materials, techniques for creating 18 projects—mirthful panorama of American characters. 105 illustrations. 80pp. 8⅜ × 11. 24135-1 Pa. $2.50

300 ART NOUVEAU DESIGNS AND MOTIFS IN FULL COLOR, C.B. Grafton. 44 full-page plates display swirling lines and muted colors typical of Art Nouveau. Borders, frames, panels, cartouches, dingbats, etc. 48pp. 9⅜ × 12¼. 24354-0 Pa. $6.00

SELF-WORKING CARD TRICKS, Karl Fulves. Editor of *Pallbearer* offers 72 tricks that work automatically through nature of card deck. No sleight of hand needed. Often spectacular. 42 illustrations. 113pp. 5⅜ × 8½. 23334-0 Pa. $2.25

CUT AND ASSEMBLE A WESTERN FRONTIER TOWN, Edmund V. Gillon, Jr. Ten authentic full-color buildings on heavy cardboard stock in H-O scale. Sheriff's Office and Jail, Saloon, Wells Fargo, Opera House, others. 48pp. 9¼ × 12¼. 23736-2 Pa. $3.95

CUT AND ASSEMBLE AN EARLY NEW ENGLAND VILLAGE, Edmund V. Gillon, Jr. Printed in full color on heavy cardboard stock. 12 authentic buildings in H-O scale: Adams home in Quincy, Mass., Oliver Wight house in Sturbridge, smithy, store, church, others. 48pp. 9¼ × 12¼. 23536-X Pa. $3.95

THE TALE OF TWO BAD MICE, Beatrix Potter. Tom Thumb and Hunca Munca squeeze out of their hole and go exploring. 27 full-color Potter illustrations. 59pp. 4¼ × 5½. (Available in U.S. only) 23065-1 Pa. $1.50

CARVING FIGURE CARICATURES IN THE OZARK STYLE, Harold L. Enlow. Instructions and illustrations for ten delightful projects, plus general carving instructions. 22 drawings and 47 photographs altogether. 39pp. 8⅜ × 11. 23151-8 Pa. $2.50

A TREASURY OF FLOWER DESIGNS FOR ARTISTS, EMBROIDERERS AND CRAFTSMEN, Susan Gaber. 100 garden favorites lushly rendered by artist for artists, craftsmen, needleworkers. Many form frames, borders. 80pp. 8¼ × 11. 24096-7 Pa. $3.50

CUT & ASSEMBLE A TOY THEATER/THE NUTCRACKER BALLET, Tom Tierney. Model of a complete, full-color production of Tchaikovsky's classic. 6 backdrops, dozens of characters, familiar dance sequences. 32pp. 9⅜ × 12¼. 24194-7 Pa. $4.50

ANIMALS: 1,419 COPYRIGHT-FREE ILLUSTRATIONS OF MAMMALS, BIRDS, FISH, INSECTS, ETC., edited by Jim Harter. Clear wood engravings present, in extremely lifelike poses, over 1,000 species of animals. 284pp. 9 × 12. 23766-4 Pa. $8.95

MORE HAND SHADOWS, Henry Bursill. For those at their 'finger ends," 16 more effects—Shakespeare, a hare, a squirrel, Mr. Punch, and twelve more—each explained by a full-page illustration. Considerable period charm. 30pp. 6½ × 9¼. 21384-6 Pa. $1.95

KEYBOARD WORKS FOR SOLO INSTRUMENTS, G.F. Handel. 35 neglected works from Handel's vast oeuvre, originally jotted down as improvisations. Includes Eight Great Suites, others. New sequence. 174pp. 9⅜ × 12¼.
24338-9 Pa. $7.50

AMERICAN LEAGUE BASEBALL CARD CLASSICS, Bert Randolph Sugar. 82 stars from 1900s to 60s on facsimile cards. Ruth, Cobb, Mantle, Williams, plus advertising, info, no duplications. Perforated, detachable. 16pp. 8¼ × 11.
24286-2 Pa. $2.95

A TREASURY OF CHARTED DESIGNS FOR NEEDLEWORKERS, Georgia Gorham and Jeanne Warth. 141 charted designs: owl, cat with yarn, tulips, piano, spinning wheel, covered bridge, Victorian house and many others. 48pp. 8¼ × 11.
23558-0 Pa. $1.95

DANISH FLORAL CHARTED DESIGNS, Gerda Bengtsson. Exquisite collection of over 40 different florals: anemone, Iceland poppy, wild fruit, pansies, many others. 45 illustrations. 48pp. 8¼ × 11. 23957-8 Pa. $1.75

OLD PHILADELPHIA IN EARLY PHOTOGRAPHS 1839-1914, Robert F. Looney. 215 photographs: panoramas, street scenes, landmarks, President-elect Lincoln's visit, 1876 Centennial Exposition, much more. 230pp. 8⅜ × 11¼.
23345-6 Pa. $9.95

PRELUDE TO MATHEMATICS, W.W. Sawyer. Noted mathematician's lively, stimulating account of non-Euclidean geometry, matrices, determinants, group theory, other topics. Emphasis on novel, striking aspects. 224pp. 5⅜ × 8½.
24401-6 Pa. $4.50

ADVENTURES WITH A MICROSCOPE, Richard Headstrom. 59 adventures with clothing fibers, protozoa, ferns and lichens, roots and leaves, much more. 142 illustrations. 232pp. 5⅜ × 8½. 23471-1 Pa. $3.50

IDENTIFYING ANIMAL TRACKS: MAMMALS, BIRDS, AND OTHER ANIMALS OF THE EASTERN UNITED STATES, Richard Headstrom. For hunters, naturalists, scouts, nature-lovers. Diagrams of tracks, tips on identification. 128pp. 5⅜ × 8. 24442-3 Pa. $3.50

VICTORIAN FASHIONS AND COSTUMES FROM HARPER'S BAZAR, 1867-1898, edited by Stella Blum. Day costumes, evening wear, sports clothes, shoes, hats, other accessories in over 1,000 detailed engravings. 320pp. 9⅜ × 12¼.
22990-4 Pa. $9.95

EVERYDAY FASHIONS OF THE TWENTIES AS PICTURED IN SEARS AND OTHER CATALOGS, edited by Stella Blum. Actual dress of the Roaring Twenties, with text by Stella Blum. Over 750 illustrations, captions. 156pp. 9 × 12.
24134-3 Pa. $7.95

HALL OF FAME BASEBALL CARDS, edited by Bert Randolph Sugar. Cy Young, Ted Williams, Lou Gehrig, and many other Hall of Fame greats on 92 full-color, detachable reprints of early baseball cards. No duplication of cards with *Classic Baseball Cards*. 16pp. 8¼ × 11. 23624-2 Pa. $2.95

THE ART OF HAND LETTERING, Helm Wotzkow. Course in hand lettering, Roman, Gothic, Italic, Block, Script. Tools, proportions, optical aspects, individual variation. Very quality conscious. Hundreds of specimens. 320pp. 5⅜ × 8½.
21797-3 Pa. $4.95

TOLL HOUSE TRIED AND TRUE RECIPES, Ruth Graves Wakefield. Popovers, veal and ham loaf, baked beans, much more from the famous Mass. restaurant. Nearly 700 recipes. 376pp. 5⅜ × 8½. 23560-2 Pa. $4.95

FAVORITE CHRISTMAS CAROLS, selected and arranged by Charles J.F. Cofone. Title, music, first verse and refrain of 34 traditional carols in handsome calligraphy; also subsequent verses and other information in type. 79pp. 8⅜ × 11. 20445-6 Pa. $3.00

CAMERA WORK: A PICTORIAL GUIDE, Alfred Stieglitz. All 559 illustrations from most important periodical in history of art photography. Reduced in size but still clear, in strict chronological order, with complete captions. 176pp. 8⅜ × 11¼. 23591-2 Pa. $6.95

FAVORITE SONGS OF THE NINETIES, edited by Robert Fremont. 88 favorites: "Ta-Ra-Ra-Boom-De-Aye," "The Band Played On," "Bird in a Gilded Cage," etc. 401pp. 9 × 12. 21536-9 Pa. $10.95

STRING FIGURES AND HOW TO MAKE THEM, Caroline F. Jayne. Fullest, clearest instructions on string figures from around world: Eskimo, Navajo, Lapp, Europe, more. Cat's cradle, moving spear, lightning, stars. 950 illustrations. 407pp. 5⅜ × 8½. 20152-X Pa. $4.95

LIFE IN ANCIENT EGYPT, Adolf Erman. Detailed older account, with much not in more recent books: domestic life, religion, magic, medicine, commerce, and whatever else needed for complete picture. Many illustrations. 597pp. 5⅜ × 8½. 22632-8 Pa. $7.95

ANCIENT EGYPT: ITS CULTURE AND HISTORY, J.E. Manchip White. From pre-dynastics through Ptolemies: scoiety, history, political structure, religion, daily life, literature, cultural heritage. 48 plates. 217pp. 5⅜ × 8½. (EBE) 22548-8 Pa. $4.95

KEPT IN THE DARK, Anthony Trollope. Unusual short novel about Victorian morality and abnormal psychology by the great English author. Probably the first American publication. Frontispiece by Sir John Millais. 92pp. 6½ × 9¼. 23609-9 Pa. $2.95

MAN AND WIFE, Wilkie Collins. Nineteenth-century master launches an attack on out-moded Scottish marital laws and Victorian cult of athleticism. Artfully plotted. 35 illustrations. 239pp. 6⅛ × 9¼. 24451-2 Pa. $5.95

RELATIVITY AND COMMON SENSE, Herman Bondi. Radically reoriented presentation of Einstein's Special Theory and one of most valuable popular accounts available. 60 illustrations. 177pp. 5⅜ × 8. (EUK) 24021-5 Pa. $3.50

THE EGYPTIAN BOOK OF THE DEAD, E.A. Wallis Budge. Complete reproduction of Ani's papyrus, finest ever found. Full hieroglyphic text, interlinear transliteration, word-for-word translation, smooth translation. 533pp. 6½ × 9¼. (USO) 21866-X Pa. $8.50

COUNTRY AND SUBURBAN HOMES OF THE PRAIRIE SCHOOL PERIOD, H.V. von Holst. Over 400 photographs floor plans, elevations, detailed drawings (exteriors and interiors) for over 100 structures. Text. Important primary source. 128pp. 8⅜ × 11¼. 24373-7 Pa. $5.95